Wachusett:

How Boston's 19th Century Quest for Water
Changed Four Towns and a Way of Life

By Eamon McCarthy Earls

ISBN-978-0-9825485-1-6

Published by Via Appia Press, Franklin, Massachusetts, 2010

www.viaappiapress.com
McCarthy Earls, Eamon

This book is dedicated to the members of the Clinton Historical Society, past, present and future.

Acknowledgements

This book would not have been possible without the help of a wide range of people. In particular, I wish to thank Terrance Ingano, the superintendent of the Clinton schools and a dedicated historian of Clinton, along with the other members of the Clinton Historical Society, whose photo collections and local knowledge proved vital. Quietly helpful, my appreciation goes out to the staffs of the Bigelow Free Public Library, the Beaman Memorial Public Library in West Boylston, the Boylston Public Library, the Leominster Public Library, the Worcester Public Library, the microfilm department at the Boston Public Library, and the Massachusetts State Archive. My special thanks are offered to Sean Fisher, the Massachusetts Department of Conservation and Recreation archivist, who revealed a hidden world of reports, letters, and insights into the Wachusett Reservoir. I also wish to thank Gordon Lankton, Chairman of Nypro, for his support and advice, as well as E. Russell and Dorothy Grady, who provided snapshots, memories of life overlooking the reservoir, and the only known film taken of the dismantling of the Clinton viaduct. In addition to all of these individuals and organizations, I wish to thank Dr. Robert Allison, chairman of the history department of Suffolk University, for reviewing the manuscript and offering brief comments for the cover.

Many other people deserve note for their indirect contributions. In particular, more than a century ago, a handful of photographers, working for the Commonwealth, preserved the construction of the Wachusett Reservoir and its impact in painstaking detail. These images are a priceless asset.

Table of Contents

Introduction

As you drive across the "Apple Country" of Massachusetts along State Routes 12, 62, 70, 110 or 140, there is a good chance that you will pass a body of water bordered by orderly stands of coniferous trees, interspersed here and there with native deciduous trees. Furthermore, if you should drive between Clinton and West Boylston, down the southern shoreline, you will pass an immense dam spanning the valley above the Nashua River, which makes it obvious that this body of water is in fact a man-made reservoir. But that is probably as far as your train of thought will go. Very few people, whether passing through or living in one of the four towns that border it, realize that this reservoir -- it is called Wachusett, a name it shares with the nearby mountain -- was once one of the world's largest, and for two generations the main supply of Boston's drinking water. That impressive stone dam also has a claim to fame, like the reservoir it was once amongst the largest of its kind.

Today, more than a century after its creation, the Wachusett Reservoir is still impressive and beautiful. It is surrounded by New England towns steeped in history, and still leading the curve -- Clinton, for instance, is a world leader in plastics. However, unlike the Quabbin Reservoir further west, in central Massachusetts, no one has ever published a complete history of the Wachusett Reservoir and the complex story of its development. This book is the first.

The Roots of the Wachusett Reservoir

The story of the Wachusett Reservoir actually begins almost three hundred years before it was even conceived. Starting with the arrival of the Pilgrims in 1620s, English settlers began to migrate to Massachusetts in ever increasing numbers. The Puritans, a strict religious sect (who were in fact different from their Plymouth neighbors) arrived first in what is now Salem in 1626 and then in Boston in 1630. At the time, the Shawmut Peninsula, where Boston would take shape, was surrounded by salt marshes. The settlement that the Puritans created there was connected to the mainland only by a narrow isthmus, helping to ensure that Boston was one of the safest settlements in the English colonies. Sequestered from Indian attacks, the settlement grew rapidly as the capital of the theocratic Massachusetts Bay Colony.

Although Boston was soon the largest settlement in the area, it still had only a few thousand people, at most. Thus, the limited supplies of fresh water on the little peninsula were not too much of a problem. Drinking water was gathered from a well in what is today Boston Common, or gathered from rainfall, in cisterns. The only attempt at an organized supply system occurred in 1652, when a neighborhood collaborated to build a four meter long reservoir, to store water carried from their different wells.

Meanwhile, thousands of Bay Colony settlers had begun to move inland up the Charles River to Watertown, coincident with the settlement of Boston, reaching the valley of the Sudbury River a few years later in 1639. By 1643, the Puritans had launched themselves several miles further to the north and west, making their first homes along the lower reaches of the long and winding Nashua River in Lancaster. It was truly a pioneer existence. Indeed, in King Phillips War (1675-1676) and subsequent "French and Indian" wars, the existence of Lancaster and other frontier communities proved to be very tenuous.

A century later, though, the Bay Colony and the region of the Nashua around Lancaster were secure and prosperous. Homesteaders grew rye, and much of the countryside was covered in rippling fields of grain. The hills of the region were also suited to growing apples, and orchards gave color then, as they do today, to towns like Bolton, Boylston and Sterling. It seems to have been an idyllic era: the Revolution was mostly experienced at a distance; Shays' Rebellion of

1786 to 1787, centered further to the west, did not directly affect the region; and Samuel Slater and his successors had not yet brought the sounds of industry to this quiet part of the Commonwealth.

Back in the east, with the British gone, Boston was triumphal after the American victory in the Revolution – a center for politics, education, and commerce, with a 1790 population of slightly more than 18,000. Although this seems tiny today, it was sufficient to rank Boston among the major American metropolises along with Philadelphia, New York, and Charleston, and the future promised more growth. Even if inadequate drinking water supplies had not yet become a crisis, that year the city did select its first reservoir: Jamaica Pond.

This modest body of water, today a component of the "Emerald Necklace" of parks, was a kettle hole pond* in the farming community of Jamaica Plain. It must have seemed a great step forward in meeting the city's water demands. To accomplish this linkage between city and suburb an Aqueduct Corporation was formed and in 1795, four separate systems of wooden pipes were built to carry water from the pond, by gravity, into Boston. However, initially, this new blessing was reserved for fighting fires and for a few wealthy land owners.

But growth continued. With a population of 50,000, Boston was officially incorporated as a city in 1822, and that called for a new outlook. Boston's mayor, Josiah Quincy, had a committee created to study future water supplies for the city. Not surprisingly, the committee recommended that the city secure additional water supplies, and, in 1825, a Professor Treadwell proposed the use of Spot Pond, to the north of the city.

**A kettle hole is a geographic feature caused by the retreat of glaciers. As a glacier retreats, large chunks of ice break away, becoming partly buried in the soil. The great blocks of ice left behind like glacial calling cards, melt, forming craters which then fill with water. These 'kettle-hole' ponds tend to be small, but quite deep. Examples include Jamaica Pond and Walden Pond. For example, despite covering only 61 acres of land, Walden Pond is 102 feet deep; Jamaica Pond is 53 feet deep.*

The city chose not to act on the issue, as there was a dispute regarding whether or not the water should be controlled by the city itself or by a private company. The matter could not be ignored indefinitely. The year 1832 saw a second committee formed to examine the need for public water supply. Led by engineering notable Loami Baldwin, Jr., (who helped his father build the Middlesex Canal between Boston and Lowell), the committee produced *A Report on Introducing Pure Water into the City of Boston* in 1834, which examined the potential of surface water resources near the city including Spot Pond, to the north, which this time was dismissed as too small, and the Charles River itself, which was not viewed as sufficiently pure. However, Baldwin and his colleagues found that water from Framingham, or Long Pond in Natick, to the west, could be piped into the city and that the quality and volume were ideal. In urging this step, the committee cited the increasing contamination of public wells in Boston -- as many as a quarter of the wells had water considered unsafe for human consumption. Furthermore, the report showed that salt water springs were often intermingled with fresh in the Boston area, making it unlikely that subsurface water could be a reliable source for the young metropolis. But, once again, the city did nothing.

In the meantime, in 1836, the General Court, flying in the face of Baldwin's advice, granted a group calling itself the Boston Hydraulic Company the use of Spot Pond and the Mystic Lakes to serve the city. Although the source of water was being given over to the control of a private company, Boston was given the right to buy one third of the stock. The lethargic (or perhaps prudent) city council decided not to purchase the available stock, and despite or because of this action the Hydraulic Company folded.

A third committee was formed in 1837, which promptly deadlocked regarding which water source to pursue. A few years later, another water bill died in the legislature and in 1843, yet another unsuccessful attempt was made to launch a private water company. Finally, in 1844, that perennial question of how to solve the city's water problems was put to a referendum. A decisive majority of voters favored Long Pond over Spot Pond. That political momentum led to another sequence of bills in the legislature, more hard lobbying on both sides, and finally, the passage of enabling legislation in 1846. This time, the project was put under the management of three commissioners and funded through the sale of bonds.

By raising the water level of 600 acre Long Pond in Natick, Lake Cochituate, 154 feet above sea level, was born. The project ended up costing Boston in the range of four million dollars. A small intermediate facility, the Brookline Reservoir, was built in that town to temporarily store the water on its 14-mile trip to the city. To celebrate the costly venture, which had been conducted under the engineering guidance of James Fowle Baldwin, younger brother of Loami (who had died in 1838), the city held a water celebration at the Frog Pond on Boston Common on October 25, 1848. Fifty thousand people watched as a valve was opened sending a jet of fresh water 80 feet into the air.

At last, with Lake Cochituate, which served as Boston's primary water source for two decades (and remained an integral part of the system for more than a century), Boston had the beginnings of a modern water system. But growth – and water demand – proved to be immense. In short order, the Upper Mystic Lakes were secured as additional sources in the late 1860s, and the first portion of the Chestnut Hill reservoir, a new storage facility, was built in 1870. In 1872, continuing the march west from Lake Cochituate, the Sudbury River Act was passed allowing Boston to collect water from the river valley of that name in Framingham. Three reservoirs and an aqueduct were finished by 1880. Then, in 1894 – the Hopkinton Reservoir was completed for just over $900,000. A year later, Ashland Reservoir was added nearby.

But these steps were stopgaps. Drought, in the early 1890s and a continually growing population caused heightened concerns about the city's water. Boston was expanding rapidly, not only in population, but in land area. The present day Back Bay and Fenway had long since been filled in, dramatically increasing the land area of the city, while the city fathers repeatedly enlarged the city through annexation of other municipalities such as Roxbury in 1867, Dorchester in 1869, West Roxbury, and Brighton and Charlestown in 1873. (Hyde Park was not annexed until 1912.) On average, Boston was growing by seventy thousand people every ten years. Its population in 1895 was almost half a million and nearly a million people lived in the Greater Boston area. The high growth rate alone made more water a necessity, but an additional problem was the unregulated use of water -- many people let their water run day and night throughout the winter to prevent the pipes from freezing.

This time, in 1893, the State Board of Health was charged with looking into the water needs of the metropolis. At this time the board employed several individuals who would become some of the most important figures in Boston's water odyssey. They included Frederic Stearns, who worked out a plan for selecting city water supplies, and Joseph Davis, who worked on both New York City and Boston's water projects. The third, and perhaps the most famous, was Xanthus Henry Goodnough, who was affiliated with Massachusetts water projects for 44 years.*

The Board of Health considered many possible locations for a new reservoir. Proposals included the Merrimack and Charles Rivers (both judged too filthy for the purpose), as well as Lake Sebago in Maine or Lake Winnipesaukee in New Hampshire, although both of these were very far from Boston. That was when it became clear, at least to those on the committee, that the easiest option was to dam the upper Nashua River, which was still relatively clean, relatively close, and offered sufficient flow to slake the city's thirst.

But where, and at what cost? The area for a hypothetical dam would probably be somewhere near Clinton, Massachusetts, in the natural valley of the Nashua. In fact, this site had been the location for the ancient Pleistocene Lake Nashua, formed as the glaciers that covered New England continued their retreat. In that unfamiliar prehistoric landscape, geologists speculate that the glacial ice front probably stood in the vicinity of Clinton, and the outlet of Lake Nashua, which the glacier blocked, was via South Clinton and from there into the valley of the Assabet River, which in turn flowed to the ancient Lake Sudbury.

Notwithstanding this geologic precedent for an eastward diversion of water, with remarkable nonchalance, the Board of Health mentioned the consequences of a dam for towns such as Clinton, Boylston and West Boylston, if the project were to go ahead: "It does not appear to us to be a very important objection to our plan that certain mill sites will be 80 feet below the surface of the basin..."

**Goodnough, a Harvard graduate, though not trained as an engineer, had begun his career working for the railroads in the far west. When he was in Massachusetts he frequently fished in the Western part of the state. Some of the locations where he fished would later become the sites for city reservoirs. He later became a chief advocate for creation of the Quabbin Reservoir.*

"…Nor that the homes of many industrious people dependent upon these mills for their living will also be submerged…."

The Board of Health submitted its report in February 1895. It passed, and at the same time, the Metropolitan Water Board was created to oversee the project, by far the largest yet contemplated on behalf of the city, and to manage it after completion. It was Goodnough who would oversee the building of the Wachusett. Of course, the plans had scarcely been made public when the towns that would be affected began to clamor for compensation – an effort that took years to reach its conclusion.

The scale of the project was immense, far beyond anything attempted before in Massachusetts with the possible exception of the Hoosac Tunnel, which breached the Berkshires and linked Massachusetts by rail with the west. Among reservoirs it also ranked at the time with the largest built or being built anywhere.

A Valley to be Dammed

Although the creation of the Quabbin Reservoir has received some attention in Massachusetts, particularly because of the four Swift River Valley towns demolished, and disincorporated to create it, the Wachusett was on nearly the same heroic scale, yet is scarcely known or appreciated. Despite the many challenges of scale and cost, Wachusett Reservoir was built rapidly over a 13 year period, with most of the construction conducted between 1895 and 1904, when the reservoir began to fill.

Although none of the Nashua River Valley towns were completely submerged by the Wachusett, and none lost their municipal identity, the dislocation of people and industry was comparable to that suffered by the Swift River Valley towns little more than a generation later. Some 360 homes, four churches, six miles of railroad tracks, eight schools, six major mills, and 19 miles of roadway were removed. Clinton, then as now the largest town in population, (in 1895 it had 11,497 people) was the least affected, as very few people had to be relocated (however, the Wachusett dam would become a major mark on the Clinton landscape). Rural Sterling, along the northern edge of the reservoir, also suffered only minimally from the project.

West Boylston, which had 3,000 citizens, five churches, six mills, and ten schools, was the closest rival to Clinton. West Boylston, although it lost only 17 percent of its land, ended with half of its population displaced, the manufacturing village of Oakdale largely disappeared, and a total of 2,000 acres of farmland submerged or acquired as buffer lands.

But statistics alone don't tell the story. For more than 200 years, the towns had grown, forging individual identities but sharing a life with a powerful and valuable river. The Nashua River extends for a total of 56 miles, cutting through northern Worcester County, and then eventually into New Hampshire. The name Nashua is derived from a Native American word, 'Nashaway,' meaning the 'River with the Pebbled Bottom.' The southern branch, or South Nashua River, rises in West Boylston at the confluence of several streams, especially the Stillwater and Quinapoxet Rivers. Originating to the north, across a range of hills, the North Nashua River flows southeast from its beginnings in Fitchburg, meeting up with the South Nashua River at a junction in southeastern Lancaster. Continuing north of this meeting place the river flows seaward, weaving through the countryside, meeting the Squannacook River, between Shirley, Ayer, and Pepperell, and then traveling further north into New Hampshire, finally joining the Merrimack River in the city of Nashua.

The story of the reservoir towns begins with Lancaster, Massachusetts, the oldest settled town in Worcester County and at that time one of the true frontier settlements of the Bay Colony. Lancaster was the "mother town" that spawned many of the towns in north-central Massachusetts. Its offspring include Stow, Bolton, Clinton, Hudson, Marlborough, Leominster, and Berlin. During the King Philip's War of 1675-76, Lancaster was put to the torch and many of the settlers were killed, including most of the family of a woman named Mary Rowlandson, who was taken captive by the Indians. Her account of the event, which became a kind of 17th century best seller, captures the terrors of the Massachusetts frontier:

"On the tenth of February 1675, came the Indians with great numbers upon Lancaster: their first coming was about sunrising; hearing the noise of some guns, we looked out; several houses were burning, and the smoke ascending to heaven. There were five persons taken in one house; the father, and the mother and a sucking child, they knocked on the head... Now is the dreadful hour come, that I have often heard of (in time of war, as it was the case of others), but now mine eyes see it. Some in our house were fighting for their lives, others wallowing in their blood, the house on fire over our heads, and the bloody heathen ready to knock us on the head, if we stirred out."

Rowlandson spent several months in captivity, and according to a mixture of legend and history, was later released at Redemption Rock, not far from Mount Wachusett. After her release, Mrs. Rowlandson's story made her one of the most important female authors of her day, and arguably, the first substantial writer in American history. Her *Narrative of Captivity and Restoration,* set a precedent for an entire non-fiction genre, known as 'captivity narratives' about kidnappings by Indians, whereby the victim is restored to civilization thanks to their faith in god.

Although Lancaster survived this baptism by fire, the losses of King Phillips' War retarded growth for years. But when settlers returned, Lancaster eventually prospered enough to give birth to its many daughter towns. Over three centuries, Lancaster's immense territory, dwindled to its present size. The last town to break away was Clinton, in 1850.

Transcendentalist and prolific travel writer, Henry David Thoreau, recalled Lancaster and the region of the Nashua, in his travel journal, *A Walk to Wachusett,* which recounted a walking trip in 1842. With his companion, Thoreau paused in the neighborhood of Lancaster. "Before noon we reached the highlands overlooking the valley of Lancaster, (affording the first fair and open prospect into the west,) and there, on the top of a hill, in the shade of some oaks, near to where a spring bubbled out of a leaden pipe, we rested during the heat of the day, reading Virgil, and enjoying the scenery…The lay of the land hereabouts is well worthy of the attention of the traveler. The hill on which we were resting made part of an extensive range, running from southwest to northeast, across the country, and separating the waters of the Nashua from those of the Concord, whose banks we had left in the morning."

Thoreau also commented on the river: "The descent into the valley on the Nashua side, is by far the most sudden; and a couple of miles brought us to the southern branch of the Nashua, a shallow but rapid stream, flowing between high and gravelly banks."

Thoreau then passed through Sterling on his way to Mt. Wachusett, and recorded his impression of the village. "Passing through Sterling, we reached the banks of the Stillwater, in the western part of the town, at evening, where is a small village collected. We fancied that there was already a certain western look about this place, a smell of pines and roar of water, recently confined by dams, belying its name, which were exceedingly grateful. When the first inroad has been made, a few acres leveled, and a few houses erected, the forest looks wilder than ever…This village had, as yet, no post-office, nor any settled name. In the small village we entered, the villagers gazed after us, with a complacent, almost compassionate look, as if we were just making our debut in the world at a late hour. 'Nevertheless,' did they seem to say, 'come and study us, and learn men and manners.'"

With approximately 800 families resident in the 1760s, most from English backgrounds, Sterling was the first of the future reservoir towns to leave Lancaster. It was incorporated in 1781. Before it was dubbed, Sterling, the area was called Chocksett, an Algonquin word for 'land of the foxes.' The new name the town took from General William "Lord Stirling" Alexander,* an American general of Scottish descent who briefly commanded the Continental army during the Revolution. So fondly was he regarded by his men, some of whom resided in Chocksett, that they named the new municipality in his honor. Later, in remembrance of the town's namesake, the Marquis de Lafayette visited on his 1824 American tour. Isaac Goodwin, Esq., chairman of the board of selectman, addressed the general and those assembled, as follows:

"The name of this town associates with it the recollections of another transatlantic hero, who, like yourself, sir, felt a sympathy from father's wrongs, and whose sword was unsheathed for their redress. Lord Sterling [sic], the gallant and the generous, now sleeps in dust, but the memory of America's benefactors will survive the decay of time. The multitudes that hail your march through this part of our country are not the assemblages of idle crowds, seeking to gratify a morbid curiosity, but, sir, the men around you are the independent possessors of their fields, and the defenders of their homes. From hoary age to lisping childhood, our whole population are eager in contributing deserved honors to the companion of Washington, the benefactor of our country, and the friend of mankind."

Unlike the other reservoir-side towns, Sterling remained primarily agricultural, although it developed some local cottage industries including woodworking, needle-making, hat making, and clocks. Cottage industries and farming aside, Sterling had a few small scale factories. West Sterling Pottery produced bean pots and other items, and the Buck Chair factory in next-door, Princeton, Massachusetts employed many Sterling residents. Although the town began to attract some Irish, Lithuanian and Polish immigrants, it still had a smaller population than West Boylston or Clinton.

The town had one particularly famous native in the 19th century, Mary Sawyer, often credited as the inspiration for the verse, "Mary Had A Little Lamb." Indeed, Sterling always seems to make a name for itself indirectly. The *Old Farmer's Almanac* was first put together in Sterling by Robert B. Thomas, the founder (who had been a school master in Boylston). Another local, Ebenezer Butterick, and his wife Ellen Augusta Pollard Butterick, invented "graded" patterns for sewing different size garments.

**Alexander claimed the vacant title of Earl of Stirling, in the Peerage of Scotland.*

The business grew so fast that the couple relocated to New York City and then opened offices across the country. Later, Butterick, started magazines to promote his products. If these entrepreneurs had stayed in Sterling, perhaps the town might have grown into a city in its own right!

But then, as now, the town's natural beauty seems to have been reckoned enough. For decades, the Waushacum ponds in Sterling were home to a popular vacation destination. Twenty-two tents were erected in 1852, at the behest of the Campground Association, a Methodist organization, which eventually built up cottages and a meeting hall at West Waushacum Pond that operated well into the 20th century. The popularity of Waushacum Park and Campground was further enhanced by the arrival of the railroad. The New York, New Haven & Hartford Railway had built its rail line along a causeway across West Waushacum Pond, creating a smaller pond, the Quag. "Progress" also came on the water. A steam yacht, *Zephyr* was built as an attraction for the park, and each summer it chugged around the pond laden with fashionable passengers.

A similar story of growth, albeit eventually with more industry, is told of Boylston, which was settled in 1705. Members of the Keyes family were amongst the first settlers. A weaver by trade, John Keyes settled with his wife and his ten children on the land that would become Boylston. At the time of the American Revolution, Boylston, with five hundred citizens, was the site of a confrontation between patriotic radicals, and conservative royalists, when Rev. Ebenezer Morse took the fight to preserve support for the crown to his pulpit. Alas, the well educated minister's ideas were no longer aligned with the thinking of much of his flock, and he was soon forced out of his position. Nonetheless, by one estimate, roughly half of the Boylston population was initially opposed to independence and did their best to hinder the Revolution. Along with other 'Tories' they hatched a plot to flood the colonies with counterfeit money. In fact, one of the centers of this colonial counterfeiting ring was Bush's Tavern in Boylston. The plot failed, and the Tories were forced to either agree with those favoring independence -- or pack up and leave.

Not long after the end of the Revolution, in 1786, Boylston broke free of its parent town, Shrewsbury. Boylston today sits on the south shore of the reservoir. The town and Boylston Street in Boston are named after the same man: Ward Nicholas Boylston, a member of a prominent Boston merchant family (with a Loyalist streak).

Boylston was memorialized in the town's name when he donated money for the town to construct a public building.

But "independence" wasn't the end to internal frictions for Boylston. In the early 1830s, the Trinitarians and Unitarians were gnashing teeth over religious doctrines. After some rather nasty words were exchanged between the respective ministers of the two sects, the Trinitarian minister accused the Unitarian minister of libel and slander. In court, the Unitarian minister was found innocent, but the case served to divide the town into the 1840s.

Boylston was later called home by John B. Gough, a radical in his own right. Gough emigrated from Great Britain in the 1820s at a young age. He pursued several professions, at all of which he failed.

Living in penury, Gough became interested in the temperance movement, transforming himself into a devout teetotaler. He then began a career as a temperance speaker, traveling across North America and the British Isles attempting to dissuade others from drinking. In 1848, he built his house "Hillside," in Boylston. When he finally passed away in 1886, Gough had given as many as 9600 lectures on the evils of drink. Ironically, early in the 20th century, Boylston was reputedly home to a number of illegal stills during the Prohibition era, though supposedly most of the stills were operated by Bostonians in need of a rural setting in which to make their illegal brew.

Sawyer's Mills developed as a substantial factory village within Boylston in the 1800s. It was connected by railroad with Clinton starting in 1875. During the 1880s, because the Nashua River between Sawyer's Mills and Clinton was dammed, it was deep enough to accommodate larger boats. In fact, a small steamboat puffed up and down the river between the two towns bringing visitors to Cunningham's Grove, a local attraction.

It is interesting to note that the Nashua did not form the town line between Sterling and Boylston; the river cut through Boylston, separating a third of the town, from the other two thirds.

In 1808, the western part of Boylston divorced the town center, becoming West Boylston. It grew the staples of the region -- apples and rye -- in the early days of its existence. The people of western Boylston were far from the town center. It was a nuisance going to the opposite end of town to attend meetings so the people built a meeting house of their own. West Boylston path to separation was aided by the efforts of Ezra Beaman, a leading landowner, businessman and politician. Thus, West Boylston was born by an act of the General Court on January 30, 1808 and Beaman was chosen as the first chairman of the board of selectmen and first representative to the General Court.

The West Boylston of 1808 had 600 people, and 98 houses. Rye, oats, and other grains, amounting to three thousand bushels a year were shipped to Boston. In the early days, cider was a profitable business. There were thirty cider mills in 1820, but only one by 1885. In the later years of the 19th century, West Boylston became one of many suppliers of fresh milk to Boston. Early each morning, a train would stop on its way through town to pick up loads of milk headed for the city.

Matching these transitions in agriculture, West Boylston evolved over the century from being purely agriculture to more of a manufacturing-oriented community. Manufacturing had become of great importance thanks in large part to water power provided by the Nashua and its tributaries. Cowee's Grist Mill, a holdover from the town's agricultural past, existed up until 1900 but hundreds of people were employed at its newer neighbors such as the West Boylston Manufacturing Company in Oakdale; the Holbrook Mill on the Nashua River, which produced scythe blades and cloth; and, the Clarendon Mills, which started as Beaman Mills, and produced cotton and wool products.

Oakdale, West Boylston's manufacturing village, profited from the building of the Massachusetts Central Railroad, which began in 1870, and finished in 1880. The railroad linked Boston and Northampton, and cut through Oakdale, making the distribution of goods faster and more efficient. West Boylston's people were still mostly of English descent during the 1800s, but Irish and French-Canadian immigrants arrived in the latter half of the century.

Clinton, the southern part of Lancaster, which was known for a time as 'Clintonville,' although settled in the 1600s, was not separately incorporated until 1850. The first settler in the immediate area was John Prescott, whose descendents hung onto their homestead for three generations. John Prescott first built a grist mill and later a saw mill. These were both burned to the ground by Indians, but they prefaced Clinton's industrial future. At the outset of the 19th century, within the landscape that would become Clinton, Samuel Plant started one of the first cotton mills in America. At the same time, the Chace family started a tannery. Another local entrepreneur, James Pitt (who had sold the land for the Chace tannery) started a shingle factory. More mills opened over the first half of the 19th century, and comb making came to Clinton. For a time, Clinton was one of the largest centers of comb making in the country. Toward the end of the century, the comb factory was succeeded by a meat processing plant.

But it was the Bigelow brothers who were responsible for most of Clinton's manufacturing fame. They moved to south Lancaster in the late 1830s, and started the Clinton Company. The younger Bigelow brother, Erastus, took the company's name from the local Clinton House hotel where he had stayed in the past. The Clinton House hotel was named after New York governor, DeWitt Clinton. Intriguingly, like the town that adopted his name, Governor Clinton played a large part in another famous civil engineering project -- the Erie Canal. Early on, the Bigelows produced "coach lace," a type of decorative lace used for a variety of different purposes in addition to decorating coaches, at an old mill that they purchased for bargain price. It was the special machine invented by Erastus which allowed the company to produce coach lace for little more than two cents per yard, compared to industry prices of nearly a dollar a yard. The Bigelows moved on to start the Lancaster Quilt Company, the Bigelow Carpet Company, and the Clinton Wire Cloth Company.

With 3,000 people, and the innovative Bigelow brothers, Clinton was soon ranked as one of the top four manufacturing towns in Worcester County. The Bigelows had created a manufacturing niche by developing increasingly specialized machines for weaving wire cloth, gingham, and rugs. By the 1880s, Clinton led the world in production of these products; in fact Clinton's factories were the largest producers of carpets in the world, at the time. Bigelow carpets were found in the US Senate, the Waldorf Astoria and were even exported overseas. In fact, Clinton would continue to lead the world in carpet production until the 1930s.

During the American Civil War, Clinton was one of the first Massachusetts towns to contribute to the Union war-effort, easily overfilling its enlistment quota for troops. As it grew, Clinton attracted immigrant workers to its factories, creating a diverse community. The original settlers, who came from English backgrounds, eventually shared the town with Irish, Jewish, German,

Polish, and Greek immigrants. Clinton's many churches are testament to its religious variation. The Baptists and Unitarians were some of the first sects represented in town. Irish Catholics soon followed, holding their first masses out of doors. Poles and Greeks established their own religious communities, alongside Jews and Seventh Day Adventists. The cosmopolitan atmosphere of Clinton helped produce the first Catholic governor of Massachusetts, Leominster-born David I. Walsh, who moved to Clinton in his youth, graduated from Clinton High School in 1890, and later represented Clinton in the state legislature and became the first Irish-Catholic senator from the state.

The stage was set. The towns of the Nashua River prospered and grew in something like a pleasant isolation – satisfied with their resources, broad in their ambitions, and optimistic in their outlook. Yet to the east, the great metropolis, grown to an assertive size, contemplated its future and its unslaked thirst.

The Valley

Cowee's Grist Mill in West Boylston was owned by Edward Cowee. Notice that the land in the foreground has already been cleared, to make way for the reservoir.(C/W MARS Digital Treasures)

A view of West Boylston from across the Nashua River, with cleared land in the foreground. (C/W MARS Digital Treasures)

An image of the Odd Fellows' hall and St. Anthony's church in West Boylston. (C/W MARS Digital Treasures)

A man stands alongside his bicycle with Clarendon Mill pond and Clarendon Mill in the background. In the mid 1890s, Clarendon Mill employed close to two hundred people. (C/W MARS Digital Treasures)

"A Bird's Eye View of Clinton" from 1876, showing the town's mills, and Coachlace Pond prominently. (Library of Congress)

Decision to Build the Reservoir

America, in the early 1890s, was a country in crisis. The 'Gilded Age' of the 1880s was over, and an economic depression had emerged that was nearly as bad as the Great Depression of the 1930s. The toughest year in the economic slump was 1893, when a panic on Wall Street left many native-born Americans and new immigrants out of jobs. Clinton, Massachusetts and the surrounding towns were not as hard hit by national events as other locations, and the towns remained prosperous, though there were periodic strikes, and labor troubles. For the most part, these were just hiccups in the day-to-day existence of Clintonians, and their neighbors.

The Nashua River continued to flow, a steadying artery that brought vitality to the towns along its banks. However, this same river was eyed covetously by politicians and planners in Boston who saw in it the answer to their prayers. As a reason for Boston's thirst, the new Metropolitan Water Board later stated that, "During the twenty years since the year 1875 there have been nine years of drought, in which the rainfall during the drier six months, has been only 36 to 61 per cent [of normal rainfall.]" The Water Board continued, "Should a year like one of the latter two occur before additional sources can be made available, it would be difficult if not impossible to provide all the water needed for the cities and towns…"[served by the Water Board].

The countdown began in 1892 when the Boston Water Board advised the mayor to seek a new supply of water. On his behalf, in June of 1893, the legislature approved an expenditure of $40,000 to study three options: bringing water from Lake Winnipesaukee (spelled at the time Winnipiseogee); the Merrimack River within Massachusetts; and, the South Nashua at Clinton. When the commission studying this matter submitted its report, the last choice, Clinton, was their recommendation. The interest from Boston in building a reservoir along the Nashua apparently leaked out in the midst of the `93 slump thanks to the hard-to-hide activities of survey crews. This shocking development soon found its way to the front page of the *Clinton Daily Item*, which reported it with little effort to hide a tone of disapproval:

August 11, 1893

The Item has made references to the fact that Boston is quietly working up another scheme by which to drain the interior of the state in the interest of her water supply; the last blow was in Southboro where arrangements have been made for a big dam and extensive reservoir, with the great damage of that community.

Now, preliminary surveys are being made nearer home, in the adjacent town of West Boylston. If the trend of the surveyor's chains are an indication it seems…as if a large dam, by which an immense reservoir shall be constructed, covering a tract of land six or eight miles long including the village of Sawyer's Mills and the Clarendon Mills property in West Boylston and affecting the Oakdale privileges; while not counting the Clinton Mills, would annihilate all the privileges of the Nashua, above Clinton…

The scheme is a big one and somewhat commensurate with the future heavy demands of Boston, and would almost seem to be too gigantic in its details to even become a matter of fact; Ashland and Hopkinton by which several mill privileges were ruined, and many valuable farms submerged, and which process is now being repeated in Southboro, shows us that no scheme of this sort is too vast to be beyond the range of Boston's aspirations and abilities.

This project is one which will be closely watched by residents of this part of the country, who must be on the alert that their rights are not stolen or infringed upon by our growing metropolis. (Daily Item).

Boston's plans were clarified further later in the summer, on August 24, 1893, when the article, "The Big Reservoir," appeared on the front page.

Plans, Extent of Lake and of Damages and Prospect

The State Board of Health with their engineering staff expressed much satisfaction with the situation on the Nashua river and with the advantages which could here be had, in case Boston should decide to build the proposed big reservoir; the most feasible location for the high dam is not at the Ox-bow , as previously stated, but as low down as at Johnson's brick house, the eastern end of the dam tending toward John. F. Philbin's place and from there across D.W. Carvills Plain to the high hill, south of the C.M.R.R.[Central Massachusetts Railroad], touching the road a short distance below the railroad bridge.

At the proposed height of the dam, the water will flow to a point, in Oakdale, only five feet below the B&M railroad track-- which fast demonstrates the almost complete submergence of the Oakdale section, West Boylston and Sawyer's Mill villages and adjacent country; and the engineers say that the dam could be made, at its suggested location, even five feet higher, which would carry the water up to the Oakdale tracks, the length of the lake would be six or seven miles , extreme width of over one mile, and its depth, in the interval below Sawyer's Mills, 87 feet -- drawing from a water shed of 100 square miles -- the largest and best, so the city engineers say, of any in the state.

The present talk is that, if the scheme matures, the water will be carried by tunnel to Sudbury, as the nearest and most inexpensive method and route.

It is evident that by the conversion of this large tract of land into a lake, and the general wiping out of this section of country, an immense damage will result, not only to the territory thus inundated but also to Clinton in the loss of its own building territory as well as of adjacent country into which the town might be extended; we also lose in the fact that in the construction of such a reservoir we should be compelled to make long round-about journeys to reach points now near at hand; then there is the possibility of the danger to the Lancaster Mills' privilege, although "it is said" that the permanence and value of this privilege will be guaranteed and no interference allowed with the natural flow of the streams.

The question also arises whether by such a scheme there would be an interference with our legal rights to the waters of Waushacum.

The engineers are making similar preliminary surveys at other points in the commonwealth so that those who might be elated or depressed over possibilities will understand that the entire project; with the immense waste and damage for the interior of the state, is yet a matter of uncertainty, notwithstanding the liabilities of future or distant trouble which may seem to exist. (Daily Item).

The new reservoir slated for construction was originally to be called the "Nashua Reservoir." The aqueduct and dam associated with it were also to bear the same name. The full details of the name change are still not known, but letters between the Massachusetts Historical Society in Boston, and Henry Walcott, a member of the Metropolitan Water Board, have been preserved.

One letter, sent to Walcott, was from a man named Henry Nourse, who lived in South Lancaster (not to be confused with the South Lancaster, settlement, 'Clintonville').

"Remembering your request I have reviewed the several names attached by the Indians to prominent features of the region about the great Reservoir...I think the original orthography of the word should be restored both for historic reason and because this will avoid the appearance of partnership in the designation with a New Hampshire city which borrowed it long ago."

"The name Nashaway (now Nashua) never belonged to the main river until the red [men] wholly disappeared from the valley. From the 'Meeting of the Waters,' in Lancaster to the Merrimac the river was originally known as the Penacook. Later it was called 'Lancaster River' and 'Groton River,' and very gradually the name Nashaway crept up the 'North River' to Fitchburg and down the main stream to the Merrimac and was corrupted into Nashua. Nashaway was the phonetic spelling of the Indian name attached to the locality, to the river, and to the tribe which had its villages on the Southeast slope of Wachusett and on Waushacum lakes...The next dominant name in this neighborhood is Wachusett...If objections to the name Nashaway seem weighty enough to prevent its retention I hope the name Wachusett may be chosen for the reservoir."

In another correspondence, the choice of names was discussed further.

> *I am glad that the Board of Metropolitan Water Commissioners are considering the question of giving the name Wachusett to the large basin of water…The word 'Wachusett' is of Indian origin, and signifies near the hill or mountain. 'Wadchu' in the Indian languages means hill or mountains, and the affix 'sett' means near or in the neighborhood of…I have written somewhat in detail on this matter, even at the risk of being somewhat tedious, but I think the subject in connection with the Water Basin warrants a little prolixity.*

Boston Decides to Build

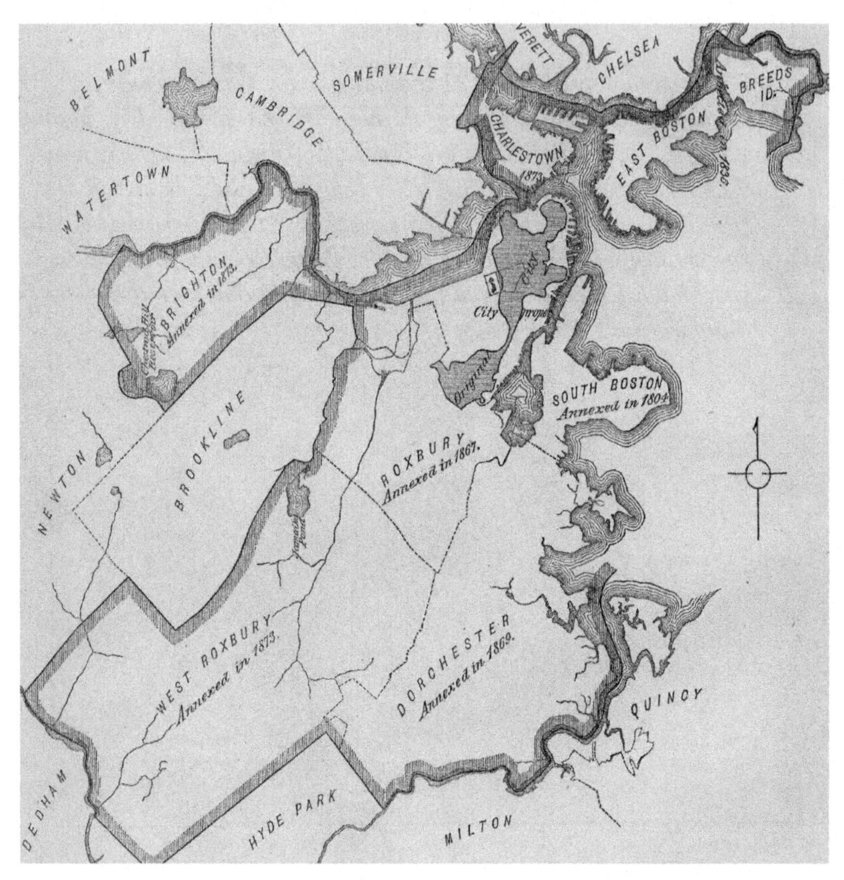

Boston's water shortage became more urgent with the annexation of new towns. Places such as Dorchester, Roxbury, and West Roxbury, were absorbed by the city and included in the metropolitan water system. This map shows Boston in 1880, with the dates of annexation included. (Perry-Castaneda Map Collection)

Before construction began, engineers carried out test borings to understand the rock on which they planned to build the dam. Advanced drilling equipment was floated on improvised rafts, such as this one. (Clinton Historical Society)

A stretch of the Nashua in Clinton. (Clinton Historical Society)

Although this photograph was not taken in the immediate vicinity of the planned Wachusett Dam, it gives an idea of what the South Branch of the Nashua River looked like in the 1890s. (Clinton Historical Society)

Construction

Once Boston reached its decision to build, there was no turning back. It would take eleven years until the project was completed, plus two more years as the reservoir reached full capacity. The bill would ultimately total 11 million dollars. Out of this, $3,700,000 would be spent to reimburse people whose homes had to be moved or demolished, and to compensate the owners of the many mills and factories that would have to close.

The summer of 1895 was a thirsty one for Boston. An extended drought dessicated the state, but its effects were most felt in Boston. With that additional inducement, the legislature passed The Metropolitan Water Act, chapter 488 of the acts of 1895 on June 5, 1895, which created the Metropolitan Water Board, chartered to not only construct the giant new "Nashua Reservoir" as it was then styled, but also to operate all the existing and planned infrastructure of the whole metropolitan system. By now, the water supply was not just for Boston itself but for a host of satellite communities including Chelsea, Everett, Malden, Medford, Newton, Somerville, Hyde Park (which was still independent of Boston), Melrose, Revere, Watertown, and Winthrop. Significantly, any other municipalities located in whole or in part within 10 air miles of the state house could, at their option, choose to join the system in the future.

One of the first steps that needed to take place was test drilling into the rock around the planned dam site, to verify the stability of the rock and the nature of the earth and rock around the site of the reservoir. This would show how deep the dam's foundation would need to be built and what would be needed for the dikes that would define much of the dam's shape.

And, almost immediately, the tide of people began to arrive. On July 30, 1895, the *Daily Item* reported some new developments at the proposed dam site and the arrival of some key personnel.

The Work Begun-Engineer Richardson Arrives in Town-

The Commission Here Today

The initial steps of the work on the Clinton dam of the Metropolitan Water Supply, were taken yesterday…when the construction of two floats about 20 by 30 was commenced on the bank of the Nashua river just above the brick house on River road.

Engineer T.F. Richardson of Boston, arrived yesterday and is registered at the Clinton House, where he will remain until he can find suitable tenement for his family.

Three carpenters are at work constructing the two big floats, each of which will be supported by twenty five empty oil barrels. When completed, which will probably be tomorrow, the floats will be towed to the site of the center of the big dam.

The engineers will arrive Thursday and borings will be commenced at once on all the territory on which the dam will stand and for several hundred feet on either side so as to give a perfect idea of the foundation upon which the immense structure will be built. The soundings or borings will be every foot one way and every 20 feet the other, so that it will require several months to complete them. Three boring machines will be used at the outlet....The work of building the big rafts is watched by large crowds of strikers and loafers. (Daily Item).

Of course, true construction would not start until the bulk of these preliminaries were complete, and it would take several months until the borings were substantially finished. With dedication, the boring teams bit into the rock at the future dam site even as cold weather moved in. By November, when one might expect work to have stopped, 60 people were employed in the day-to-day running of operations.

Although the drills were state-of-the-art technology (eight diamonds comprised each drill head), they were mounted on the crudely assembled barges or "floats," each powered by a steam boiler. Most of the boring was concluded after a few months, and preliminary work slowed as snow coated the ground. With work underway, the fact that permanent geographic changes would take place along the Nashua may have finally struck home among locals. Surprisingly, however, although it voiced concern about the reservoir project, the *Daily Item* paid as much or more attention to the strikes taking place in town, and the expensive plans to build a new town armory, rather than the reservoir. Even seemingly trivial intermunicipal disputes, such as the Clinton street railway supposedly running too fast through Lancaster, gained equal space on the front page with the gargantuan civil engineering project.

Winter brought most activities on the reservoir project to a standstill each year while warmer weather brought a rapid escalation in the size and scope of activities. The Metropolitan Water Board, which immediately took up office space in Pierce's Block, Clinton and in a satellite office in West Boylston had had only 60 employees on the project at the end of 1895, but saw its ranks swell to 1,000 workers by August 1896. Surveyors even took over one corner of Clinton's town common as a convenient spot to triangulate and measure distances.

On May 21, 1896, Hiram A. Miller was appointed to head the project while his predecessor, Richardson was allowed to focus fully on borings and other studies required for the balance of the project. In addition to borings to determine the strength of the underlying sediment and location of bedrock, scores of holes were also dug to assess the depth of the top soil so that an estimate could be made of the effort needed to remove same.

The borings for the dam consisted of two types of operations. In one, large pipes were driven into the soil and water was used to wash the contents of the pipe to the surface for analysis, thereby providing a clear picture of the nature of the underlying ground. The second type utilized the diamond tipped drill (a second diamond drill rig was added to the project in February). A total of 806 pipe borings were made in 1895 and 1896 with all being complete by June of the latter year. Thirty eight diamond drill borings were made, with a total depth of half a mile. Professor W.O. Crosby of the Massachusetts Institute of Technology was given the task of studying the results. Geologically, he found that the northwest slope of the gorge proposed for the dam was mostly metamorphic slate and schist while the opposite side of the gorge was underlain with granite – the two sections being firmly "welded" together at a midpoint. These findings were deemed to indicate a fine basis for constructing a dam.

For the building of the dike a similar process was initiated. The result showed a considerable depth of very fine, largely waterproof sand beneath the area. But to "make sure" the Water Board established an experimental facility, a 70-foot long building with a large water-tight tank within which an eight foot tall dike was constructed. One side of the dike was filled with water and, on the other side, leakage through the soil was monitored. This process gave confidence that the soil from the dike area was indeed nearly waterproof. Other filtration experiments were conducted on different soil samples as well. The smaller South dike, planned for an area with bedrock nearer the surface, mixed the techniques used for studying the dam site and the North Dike site and included driving hundreds of iron rods to ascertain the depth and location of bedrock.

To prepare the ground for the project and to provide information needed to compensate property owners, the Water Board engaged in a detailed survey of some 18 square miles and, by year's end, detailed drawings showing the land on a scale of 100 feet to the inch were under way. Indeed, 600 such drawings were completed that year. In addition, hundreds of deeds were copied, 150 pre-existing ground plans or maps were laboriously traced, and plans were made to reroute 17 miles of roads.

As to the valley's railroad (the Central Massachusetts Railroad, also known as the Central Mass., Mass. Central, or CMR), which hugged the low ground soon to be flooded, surveys were begun for multiple alternate routes to see which might be most feasible.

Meanwhile, of equal importance with the reservoir itself, was the building of an aqueduct to Boston, since the water in the Wachusett would be no use to the city, unless the Water Board could get it there. Although the reservoir itself was still in the survey and design phase, work had been proceeding on the Nashua (Wachusett) Aqueduct from the end of 1895. A major feature of the aqueduct was the great granite structure that would span the Assabet River in Northborough. In fact, it was here that the project exacted its first toll in human life. In an unfortunate incident, the embryonic reservoir claimed as its first victim Michael Holloran, a laborer from Rutland, Maine, who drowned when he fell into 15 feet of water, while 'wheeling earth.' He had been in the process of building a section of the masonry for the Roman-style aqueduct over the river.

But the spanning of the Assabet was only one element in a larger picture. No less impressive than this monumental above ground structure was the effort put into building the "rock tunnel," with a planned length of nearly two miles. Although excavation didn't commence until March 31, 1896, progress was substantial. A siding was extended from the Central Mass. Railroad and four steam boilers of 100 horsepower each were installed to power two, double Rand duplex compressors. With these aids powering their rock drills, some 300 laborers, operated through several shafts or portals, and by year end nearly half of the excavation was completed.

In some sections, where the rock was considered unstable, the rough tunnels needed to be bricked to complete the work. Beyond the extent of the Assabet crossing and the rock tunnels there were miles more for the water to travel before it linked with the existing tendrils of the metropolitan system. In the cases where the aqueduct needed to travel at or near ground level base, a ditch, semicircular in section, was excavated and filled with concrete and brick. Sections of curved iron plate, in matching semi-circular section, were placed on top and a concrete arch was cast in place, the iron plates later being removed. This additional aqueduct work employed hundreds more men.

The Water Board's plans also needed to be coordinated with the relocation of both roads and rail lines, because at some point the removal of soil and then the rising waters would cover both – making them unavailable for supporting the project.

The roads were a relatively simple task – and one that was of course important to local commerce. Eventually, the project would include a new road between Boylston Center and West Boylston, with an extension to Shrewsbury and Lancaster Street in West Boylston, and that would be complemented with a new road branching off to Clinton. The process of road building was overseen by the Water Board, but the Massachusetts Highway Commission was in charge of determining the new route to Shrewsbury from West Boylston. Most of the roads built in connection with the project were unpaved, being merely strips of gravel or dirt. However, the new road to Shrewsbury was paved with stone -- by 1890s standards, a modern highway.

On July 8, 1897, local construction companies were given a chance to bid on the building of the new road to West Boylston. The Metropolitan Water Board wanted the new road to be finished by the first of November, the same year, leaving the lucky contractor less than four months to complete the work.

The *Daily Item* reported: *"The Metropolitan board has advertised for the construction of the new road which is to supplement the present road to Boylston Center via the River road, as the latter is to be destroyed on account of being within the flow lines of the great basin.* (Daily Item).

At the end of 1897, the Metropolitan Water Board admitted in its report for the year that, "...little has been done toward the actual construction of the Wachusett Dam and Reservoir," as it was now termed. But given the immense and varied preparations required, that assessment was perhaps unfair. In fact, approximately one square mile had been cleared of wood, timber, and brush; a temporary dam had been put in place above the site of the planned permanent dam, and, a mile of the road that would replace the River Road between Boylston and Clinton was done (implying that the chosen contractor hadn't in fact met the deadline).

The temporary dam required the laying of a large water main to conduct the river water past the point of the construction project and deliver it to industrial users downstream. This cofferdam was a temporary structure that was not designed to be as sturdy as the finished dam. However, it would serve to block the river's flow, and allow workers to start construction. On July 20, 1897, the *Daily Item* reported that construction was set to start "soon."

Simultaneously, the *Daily Item* reported on other work at the dam site.

> *Pipe Line Almost Complete*
>
> *The pumps and boiler are on the scene –the trestle is built, the boilers are on the scene and pumping may begin the later part of the week... All but about 200 feet of the iron pipe which is to carry the water from above the coffer dam to the gate house of the Lancaster Mill, a distance of three-quarters of a mile, had been placed in position yesterday afternoon.* (Daily Item).

Weather didn't always cooperate. The Wachusett reservoir would be dependent on rainfall to keep its water level stable after it filled, but for the time being, the engineers hoped and prayed that dry weather would hold out. Each rain storm that was more than a fine drizzle could bring construction to a halt or even undo progress already made. This it did, all too often, such as on July 23, 1897, as the *Daily Item* reveals.

> *The rain of Thursday put another stop to the work of placing in position the supply pipe for the Lancaster Mills. Of late the officials have been greatly troubled on account of rains and high water in the river and as a*

consequence the date set for commencing the pumping has several times been postponed… (Daily Item).

The building of the cofferdam, and later the Wachusett dam, required the dismantling of an earlier dam, and the lowering of the water level in Lancaster Mill Pond.

It is understood that only one half of the Lancaster mills dam will be entirely removed but the stones along the other portion will be taken down to a point several feet lower than at present….When Lancaster Mills dam is lowered the mill pond will be left entirely dry with the exception of the narrow channel where the water will trickle down over the few remaining stones. The officials will not molest the course of the stream after it passes the site of the permanent dam. (Daily Item).

What did not remain unmolested was the real estate to be taken for the reservoir. Particularly in West Boylston, but also in Sterling, Boylston and Clinton, many houses, factories, churches, and schools would have to be torn down or removed before the reservoir was finished, and the ground cleared of plants and topsoil.

Compensation was a contentious subject. Mills and factories usually had better legal representation, but private citizens also sometimes found the means to get legal representation or work through the appeals process to boost the value of their compensation from the Commonwealth. For each mill and house, the Metropolitan Water Board assessed the structure's value and compensated the owner for its loss or removal. "The Board deemed it advisable, directly after the organization, to enter upon work necessarily preliminary to the settlement of damages which may be claimed by owners of various mill properties situated on the Nashua and Merrimac."

Mr. Albertson, the owner of the Albertson saw mill, in Boylston had his farm and mill assessed by the Commonwealth. The total compensation due to him was $17,000. This included $5,840 for the water power plant, $4,873 for the mill building and its foundation, and $951 for all other machinery. The assessment included the value of the house in which he lived, near Muddy Brook. The scrupulous nature of the assessment is testimony to the complex administrative challenges of even this phase of the project. The barn, surprisingly, was worth more than the house, valued at $1,590 and $2,249 respectively. His shed was worthless in the eyes of the government, but they included $114 to replace the hen house. Indeed, the Commonwealth seems to have accounted for just about every structure and bit of real property along several miles of the valley: icehouses, hen houses, and pastures, if seized by the state, were all recorded, and their owners recompensed.

The loss or potential loss of water power was also a matter of no small importance. In Clinton, the Lancaster Mill stood to lose a great deal. Lancaster Mill had a pre-existing dam near where the Wachusett Dam was to be built. John Freeman, a consulting engineer from Rhode Island, hired to assess the Lancaster Mill situation, wrote: "The water taken is precisely the same in quantity as that already settled for at Lowell, Lawrence, and Nashua, and the power which this water can produce, if utilized with the same degree of economy, is of course proportional to the number of feet of head and fall out of each place, but the local conditions are such that the damage will be relatively much greater at Clinton than at Nashua or Lowell; because at Clinton the injury will begin immediately and exist in its full extent on every working day of the year, while no injury will be suffered at Nashua or Lowell from the taking away of this water in the flood season."

Every mill that might be impacted by the project was similarly assessed. In West Boylston, Clarendon Mills, and the L.M. Harris Company in Oakdale, had the value of their water power assessed. Downstream, too, the impact of the planned dam had to be weighed, wherever the waters of the Nashua eventually flowed. The Commonwealth took into consideration the effect that the reservoir project would have in Groton, on Tileston & Hollingsworth Company, and in the river's namesake city, Nashua, New Hampshire, on the Nashua Manufacturing Company and the Jackson Company. Likewise, in Lowell and Lawrence, alongside the Merrimack River, the impact on the water power companies was carefully calculated.

Then there was the matter of the railroads. At the time of the reservoir's construction, three railroads operated in the towns. The Central Massachusetts Railroad line ran from Berlin in the east and through Boylston into West Boylston, meeting in Oakdale, the Worcester, Nashua, & Rochester Railroad, which cut through Sterling and Clinton. The third railway, the New York, New Haven & Hartford traced its route through Sterling, Lancaster, Clinton, Bolton, and Berlin.

Neither the NY, NH & H, nor the W, N & R needed to be relocated, although some of the NY, NH & H trackage had to undergo minor realignment or reinforcement. Thus, "On Aug. 6, 1896, an agreement was made with the New York, New Haven & Hartford Railroad by which the Metropolitan Water Board was allowed to construct a new stone arch bridge through the railroad embankment for conveying the water, and was required, in return to widen the embankment for the railroad and rip rap the slopes."

It was a different story when it came to the Central Massachusetts Railroad, which needed to be substantially rerouted.

The Massachusetts Central Railway had been formed in 1869 in an attempt to create a rail line linking Boston and Northampton. In less than thirty years of existence it had already had an interesting history. Promoters hoped it might eventually be extended as far west as the Hudson River but that goal never came within its grasp. Meanwhile, in 1880, stockholders voted to lease the railway to the Boston and Lowell Railroad, effective in 1886. However, in 1883, the railroad was reorganized with a new name -- the Central Massachusetts Railroad. Then, after only a year under the wing of the Boston and Lowell Railroad, the Central Massachusetts instead joined the Boston & Maine Railroad.

The complex question of how to reroute the railroad to the satisfaction of all concerned was to remain unsolved for some time.

But to the south, the aqueduct continued to progress rapidly. In September the aqueduct bridge spanning the Assabet – some 359 feet in length -- was declared complete, while two miles of tunneling through solid rock were finished by November. It was calculated that the completed structures, which had an average slope of only 1 foot in a mile, would be able to deliver water at the rate of 300,000,000 gallons each day. However, still to be completed was another three miles of open channel.

The year 1897 was also when planners finally grappled with the fact that the Catholic cemetery, the future site of Cemetery Island, would also have to be moved. The planners noted, "The work of removing a large number of bodies is a difficult one, but every means will be taken to effect the removal without disagreeable results, and with proper consideration for the remains of the deceased and the feelings of relatives and friends."

In 1898, while the nation geared up for its short and intense war with Spain, the Water Board seemed to have stepped up the pace of its giant project. Quarries were opened in nearby areas such as Boylston to supply building stone and a narrow-gauge (3-foot) railroad was set up by the various contractors to assist in moving stone, soil, and other material. It eventually operated on 27 miles of track, utilizing 25 small locomotives and 725 open gondola cars. Work was begun on the spillway area on the north side of the planned dam. Less grandiose but no less important, contracts were let for removal of soil over one half of the area of the reservoir. In the area of the North Dike, two cut-off trenches were dug, the larger of which had a width at the bottom of 30 feet and a depth that ranged from 30 to 60 feet. These cut-off trenches were filled with the more water-impervious soil, sometimes supplemented by lapped pilings, to protect against leakage and the danger of soil liquefaction. All together these trenches extended more than three quarters of a mile.

In the matter of the Catholic cemetery, the Water Board negotiated an agreement with the bishop of Springfield to purchase land in Lancaster and $92,000 was appropriated for removing the deceased.

After reconnoitering, at least 13 possible routes for the Central Massachusetts Railroad were offered for considered and, separately, the road between West Boylston and Clinton by way of Sterling was begun. For all its efforts – construction, compensating land owners, and engineering, the Water Board expended just over $357,000 dollars for the year.

The Water Board also completed an interceptor which carried the sewage flows of Clinton below the site of the dam to a holding area where a pumping station was to be built to feed sewage filtration beds in Lancaster.

An unanticipated issue arose in 1899, namely claims for compensation along the meadowlands that bordered the Nashua below the dam. Some farmers claimed that the reduction in flow, especially in the spring, would reduce the productivity of the land. Not having previously studied this matter, the Water Board sought the views of various experts and awaited the opportunity to study the matter over the following year.

By 1900, the Water Board had become so well established in the area that it had even constructed its own building in downtown

Clinton. Other offices were maintained in Sterling near the North Dike, in West Boylston (where sewage flows into the Quinapoxet were being studied), and on River Street in Clinton.

Increased water consumption in the metropolis and the dry season of both 1899 and 1900 weighed on planners, compelling them to let contracts for the construction of the great dam even before all design details were settled. To further speed progress, laborers were put to work directly by the Board in July to begin the process of clearing the ground for the base of the dam. On October 1, 1900 McArthur Brothers Company of Chicago, Illinois was awarded a contract for the structure valued at $1.6 million dollars with a completion date of November 1904. As then envisaged, the dam was to span an area of 1,250 feet with the main dam filling 800 feet of that space and standing some 200 feet above bedrock at its top. The spillway channel was planned to be some 1,500 feet long, emptying into the Nashua 800 feet below the dam. By mid-October, this effort was in the hands of the actual contractors for the dam rather than the Water Board, and progress was considerable while the weather allowed.

Meanwhile, the rest of the project had raced along as fast as the muscles of thousands of men and hundreds of draft animals (supplemented on longer hauls by the little steam locomotives) could permit. Up until 1900 just over 600,000 cubic yards of material had been removed from the future bottom of the reservoir. In 1900, nearly three times that amount, 1,678,346 cubic yards, were removed. But even so, the combined total was estimated to be only 34 percent of the total amount that needed to be moved! Most of the soil ended up as part of the North Dike but some was used in providing fill and embankments for roadways.

The North Dike cutoff trench, now almost complete, had reached a total length of almost two miles. The road (today designated as Route 110) between Clinton and West Boylston by way of Sterling, almost three miles in length, was completed and accepted by the county commissioners. Other roads completed ran three miles from West Boylston to Boylston and Shrewsbury; Temple Street for a mile, and Lancaster Street in West Boylston for a mile and a half.

Demolition, which had not affected many buildings initially, became the order of the day in 1901 as West Boylston saw the destruction of two mills, one hotel, 53 homes and 14 barns. Down river, the work of stripping soil continued apace, reaching 3,588,648 yards taken from an area of 4,200 acres.

While West Boylston was torn down, Clinton was built up: the first stone was laid for the dam on June 5, the most monumental challenge for the Metropolitan engineers, from among many. Those very first stones were laid under the direction of John Mercer, the foreman in charge, who was an employee of Winston Company & Locher, a subcontractor for the dam. Dissatisfied because the first stone had been laid at an unusual angle, he ordered his men to straighten it. This they did, tugging on it with cables until Mercer was satisfied. There was little official fanfare surrounding the event, but using chisels and hammers, workers broke off bits of stone to keep as souvenirs. By the end of the year the structure stood 40 feet tall, an even more impressive achievement when it is remembered that this height was at the base where the dam was at its thickest, representing 28,000 cubic yards of rock.

More so than any other part of the reservoir construction, the work on the dam was dependent on clear weather in order to succeed. As soon as the inclement winter weather rolled in, construction usually ceased. On the other hand, work sometimes took place through the late fall or on clear days during the winter. When spring arrived, the *Daily Item* noted that the accumulated snow, ice, and mud needed to be cleared away before work could begin in earnest.

Another year of hard labor brought the dam to a height of 96 feet and necessitated a partial dismantling of the temporary spillway and temporary dam.

In 1903 the temporary structures were completely removed and the river was allowed to flow through the pipes in the dam structure. By that time, an estimated total of more than 70,000 cubic yards of rock had been put in place, most of which came from a nearby quarry in Boylston owned by the state.

Now, it was truly a race as the dam builders aimed to take their edifice skyward and as the pick and shovel crews hurried to peel back the layers of soil and finish the dikes before the water rose. While another 50 buildings were dismantled or destroyed in West Boylston, a portent of the future was marked when crews began to plant pine seedlings along the margin of the reservoir site.

While not as impressive as the dam itself, contractors also completed a three-arch stone bridge over the Quinepoxet and a single-arch bridge over the Stillwater River.

The long-festering question of where and how to relocate the valley's railroad was finally settled on paper at least on April 3, 1902. The flooding of the valley rendered the traditionally most desirable and level route to the west completely untenable. Instead, the line would have to skirt the north bank of the reservoir. To make that feasible would require construction of a high viaduct over the Lancaster Mill Pond and a cut through the hill on the southern side of the valley below the dam site. The board agreed on a relocation along the northern bank of the reservoir, described as "leaving a point near the crossing of the Central Massachusetts Railroad over the New York, New Haven & Hartford Railroad in West Berlin, passes over the highway leading to Boylston, and thence proceeding northerly, follows in general the course of Berlin Street in Berlin; and, crossing the town line into Clinton, proceeds for about 1,500 feet in the same general course, then turns to the left and passes under the Clam Shell Road, [through a deep excavation known the Clam Shell Cut]; and thence by a tunnel under Wilson Street, 1,110 feet long, comes to a point near Boylston Street, which it crosses overhead; and thence is carried by a high steel viaduct, supported upon granite masonry piers, across the Nashua River just below the Wachusett Dam, and passes along the shore of the reservoir and the North Dike, to a junction with the Worcester, Nashua & Rochester Railroad, at a point 9,631 feet westerly from the present Clinton station on that railroad."

A branch line some three-quarters of a mile in length was also planned to provide for running trains to the passenger station in Clinton. The massive relocation – tunnel, tracks, viaduct removing and rebuilding roadbed -- were to cost in the range of three quarters of a million dollars, rivaling the cost of the dam itself.

Indeed, the relocation of the CMR, consisting of four parts, was an engineering achievement in its own right. A tunnel, "Not less than twenty feet high above the top of the rail, and fifteen feet wide," with, "A suitable lining of Portland cement concrete," was to be built. The rails would have to cross the Nashua River, directly above Lancaster Mill Pond. To this end, "A suitable steel viaduct, resting on a steel trestle supported by masonry piers," was to be built. Beyond the viaduct, a third task remained. A rock cut, up to 56 feet in depth, was needed through the stone on the side of Burdett Hill. Then, and only then, the tracks could be laid, and the new line would be complete.

So, as the valley was transformed under saw, pick, and shovel, and as the base of the giant new dam came into being, engineers, tunnel crews, and railroad workers created a spectacular new shortcut that knit the two halves of the CMR back into one. The viaduct was an impressive feat of engineering that would tower 133 feet over the river, and run 917 feet from end to end. The job of building the trestle viaduct was assigned to the American Bridge Company, based in New York City. Thirty men worked to build the viaduct, earning only fifty cents an hour for physically demanding and potentially fatal work. Clinton historian, Terrance Ingano records a statement from an engineer on the viaduct project. The engineer explained that bridge workers, "Have no fear, value life lightly, and have little thought of the morrow." This may have been a true statement but it seems unlikely the workers themselves were so cavalier. Working almost seventy years before the Occupational Safety and Health Administration came into being; workers had no ropes or safety harnesses to prevent falls. The daring club of viaduct workers clambered over the girders and beams of the trestle like monkeys.

Lured by the novelty of the trestle's construction (and perhaps looking for some morbid amusement) dozens of 'idlers' flocked to see the construction. Tragedy struck on at least one occasion: Henry Hendrickson, a laborer from East Boston plummeted 80 feet to his death when a winch broke. Hendrickson was the sole casualty of the trestle's construction. A much worse disaster was only narrowly avoided in another incident when five workers barely escaped with their lives. A 12 ton girder, being lifted 50 feet into the air by a derrick, fell to the ground after a cable snapped. Five men working on, or near the girder leapt to safety onto a girder that was already in place. It had been a close call!

The 12 ton girder was badly bent after its fall, and had damaged other construction material, suspending work temporarily. Despite these setbacks work soon resumed and the trestle was completed in a timely fashion, little more than a year after it was ordered. When it was completed, the trestle boasted an impressive thirty-five thousand rivets holding it together.

Working feverishly through the spring of 1903, workers then laid wooden rail ties, and put the finishing touches on the viaduct which opened on June 15, 1903. As soon as its official opening took place, passenger and freight trains began making their scheduled runs across the viaduct, which provided a splendid view of the dam, reservoir, and the little valley above the Lancaster Mills. Indeed, it seems likely that this "scenic wonder" may have contributed to the fondness with which future President, Calvin Coolidge reportedly regarded travel on the Central Massachusetts.

The new line extended a siding into Clinton, and a small turntable was built to facilitate operations.

Also in Clinton, not far from the Sterling town line, the CMR track joined up with the W, N & R track that led into Oakdale. A section of W, N & R track had needed to be moved 50 feet to accommodate reservoir construction, but the task was certainly much simpler than relocating the CMR. Oakdale station also changed locations, moving north 600 feet.

Showing just how much of a crescendo the dam and reservoir project was reaching, spending for the year 1903 reached $1,506,803. At the dam, a great deal of effort went into securing the ends of the dam and related structures to bedrock. The bed of the reservoir lost another 1,115,341 cubic yards of topsoil, removed from 621 acres of land. This put the project almost to the finish line in that regard, with an estimated 98 percent of the topsoil now gone.

The North Dike had progressed to the point where riprapping of the embankment (the placement of large stones for erosion control) could commence and the nursery for the project could boast that it had planted more than 700,000 seedlings; some 200 acres of land along the margins of the reservoir had been planted with spruce, Scotch pine, arbor vitae, tamarack, larches and other species.

Some bridges and embankments, in West Boylston were also constructed in 1904.

By 1905, the project was clearly coming toward completion with a total expenditure for the year of just over half a million dollars on the reservoir and dam, the latter being essentially complete except for finish stones at the top of the dam. The topmost part of the dam now stood 415 feet above the "city base" in Boston. On the engineering side, detailed new surveys were made of the "stripped" floor of the reservoir so as to permit exact capacity estimates to be made for each tenth of a foot in depth up to the 370 foot elevation above Boston. And, the Water Board report writer noted, "No action has been taken by either the Roman Catholic Bishop of Springfield or the St. John's Catholic Cemetery Association at Clinton toward effecting a final settlement under the…agreement."

That bland statement referred to the unpleasant chore of removing the dead bodies of several thousand former residents of the valley towns. The largest single move was St. John's Catholic cemetery in Clinton. St. John's cemetery was located on a small hill, much of which would be above the waterline when the reservoir filled. In the future, it would become known as Cemetery Island, but for the Water Board, it was a touchy subject.

The cemetery move was controversial, locally, and the Board involved the St. John's Catholic Cemetery Association and the Roman Catholic Bishop in Springfield, in negotiations. A new cemetery in Lancaster was set up, and the work of moving the bodies began. Laborers carefully dug up the cemetery, and carted off the remains to Lancaster. Unfortunately, with so many bodies and burial records that were often incomplete, it is no surprise that laborers were still extracting remains in ones and twos for years after the move. By 1902, the Board insisted that all of the bodies had finally been removed and accounted for. All told, 3,902 bodies were dug up and relocated.

However, the compensation for the Catholic's of Clinton took years to work out. Cryptic references to this topic appeared year after year in the Water Board reports. Yet, according to a story reported by Clinton historian Terrance Ingano, the problem may have been on the side of the church. It seems that the new cemetery was owned by parishioners who had formed a cemetery association when the move was decreed. This was apparently just fine with Father Patterson, the priest in charge of St. John's parish at the time. However, Patterson died before the cemetery move was completed and his successor, Father J.J. O'Keefe, found this state of affairs – a church that didn't own its own cemetery -- intolerable and, getting no support on the matter from his parishioners, refused to consecrate the grounds. Although widely admired in the parish for many of his leadership traits, this emotional issue, which dragged on for 10 years, eventually led to Patterson's reassignment by the local bishop, who promptly consecrated the grounds and accepted the long proffered compensation from the Water Board.

Meanwhile, four years and 19 days after it was begun, the Wachusett Dam was complete. Some 300,000 cubic yards of stone were used to build the dam. At the top, the dam was 22 ½ feet thick, while at its base, it was up to 185 feet thick. Once again, John Mercer presided as the final cap stone was laid in place at 10:30 in the morning on June 24, 1905. To mark the completion of the dam, Thomas Winston and C.H. Locher, co-owners of the contracting company, were present. As great an achievement as this was, for Locher the unveiling of the dam may have seemed less impressive since he was simultaneously engaged in another monumental project -- building the Sault St. Marie canal linking Lake Huron and Lake Erie.

Although the finished dam, with its dramatic granite face rising from the valley floor, was impressive, the structure was in fact even more massive than it appeared. Designed to last, the engineers had sought out the bed rock below the nominal floor of the valley to give a firm foundation. Thus, the original stone, laid by Mercer's men, rests as much as twenty to twenty five feet below ground level. Furthermore, the dam was not simply a pile of rocks. For instance, a 48 inch cast iron pipe was laid horizontally inside the dam, to carry water into the power house. Housed vertically inside the dam, was a seven foot wide brick lined 'well.' The well carried water from the reservoir vertically through the dam. Below ground level, the well met up with a 48 inch cast iron pipe that carried it into the future power house. Because the well and the pipe met to form a 90 degree angle, the water might become sluggish. To redirect the water from the well into the cast iron pipe, a deflector, built of pine wood was put in place. After the water flowed through the cast iron pipe, it would be used to power the generators, and then move through chambers below the power house.

The outer face of the dam was comprised of stones quarried in Chelmsford, Massachusetts. These stones were dressed in ashlar form. [Ashlar is a term used in masonry, to refer to stones that have been cut and formed uniformly, and used in place of bricks. Ashlar stones are not exactly alike, but in general they are about 35 centimeters tall, or slightly more than a foot.] Out of sight behind the ashlar stones were other, granite stones that had been not been carefully 'dressed.' These stones were mostly quarried from Boylston, a few miles from the dam. The Wachusett Dam gained considerable attention in the wider engineering community, and members of the Panama Canal Commission visited the site on September 27, 1905, to glean engineering knowledge that might be applicable to the huge canal under construction in Panama.

Amongst all of the shoveling, blasting, digging, and entrenching, the Water Board did devote some money to aesthetics. Arthur Shurcliff, a member of the famous Olmsted Brothers' landscape architecture firm, was in charge of landscaping around Lancaster Mill Pond in front of the dam. Shurcliff helped to design the bridge over the spillway, the fountain, and planned the layout of mulberry bushes and coniferous trees.

The completed dam was the largest *gravity dam* in the world, at the time. Because water exerts immense pressure on the dams that try to contain it, they need to be strongly constructed. Many modern dams harness the principle of the arch, laid horizontally, to hold back the pressure of the water. Gravity dams, like the Wachusett dam, by contrast, simply depend on the tremendous mass of stone to resist the pressure of the water.

Two other major structures were built to complete the impoundment of the Nashua's waters -- the North Dike and the South Dike. Both dikes, north and south, were very similar in their construction. As soil was moved, and the land sculpted to hold water, the soil was piled up in two dikes, at the edges of the reservoir. The dikes were useful both for disposing of the soil that had been removed, and preventing flooding. Although nowhere near as impressive as the Wachusett Dam, as much thought and labor went into building these earthen structures as went into building the gravity dam.

The completed North Dike was 3200 meters long, while the South Dike was somewhat smaller, only 760 meters long. According to Clinton native, Robert Pratt, writing in 1902, the maximum depth of the water at the North Dike is 65 feet, but on average 10 to 20 feet. The North Dike would rise 15 feet above the water, and its greatest thickness, where the fill extended for a great distance beyond the water, amounted to about 1900 feet.

Pratt noted that, "The greatest care has been taken to admit no vegetable matter or any matter at all except the finest earth." Engineers were worried that if any soil containing plants, or soil that had frozen, found its way into the dike, the embankment would be weakened.

Stretching 9,556 feet, a 30 foot wide cut off trench was excavated down to bed rock. A cut-off trench does exactly what its name implies -- it cuts off water. Even though the soil of the North Dike was compacted and compressed many times over, there was no guarantee that water would not seep in, causing erosion, and destabilizing the North Dike. In effect, by building the cut-off trench, and lining it with 'sheet-piling' (extending as much as 60 feet below ground level) engineers had created an underground, unseen dam. As an added precaution, a secondary cut-off trench was added. It was not as deep, but was lined similarly. As each new load of soil arrived from around the reservoir, it was laid down in six inch layers, moistened and then compressed. Where the dike would face the water, gravel was laid down, and large blocks of uncut stone were laid down to 'rip-rap' the slope and hold it in place. The rip-rapping stones laid down had many gaps between them, through which water could find its way. Their main purpose, however, was to act as a breakwater slowing the damage to the North Dike caused by waves, chop, and the perpetually lapping waters of the reservoir.

The construction of the North Dike began in 1898. The digging of trenches began in 1899 and ended in 1901. The dike was finally complete, in 1904.

In 1906, a final series of excavations for the waste channel adjacent to the dam effectively marked the end of that part of the project. The "granolithic" surface at the top of the dam and the walk to Boylston Street were completed in the early summer, followed by erection of a heavy brass fence. Permanent pipe connections to the Lancaster Mills were also completed and the area below the dam was landscaped and finished. As the year closed preparations were also being made to install turbine and generator components in the lower gate chamber and power house area at the base of the dam. Initially, the electricity generated was intended for use within the area of the reservoir itself. Also in 1906, as in Clinton, the Water Board found the effluent from parts of Sterling – namely the areas near Sterling Center and West Waushacum Pond – to be a threat to the quality of the reservoir's waters and thus, in September, completed four filtration beds to help filter the waste water.

Despite the precautions and engineering studies aimed to prevent liquefaction of the soil in the North Dike, the engineers overlooked a serious flaw; most of the soil used to build the North Dike was sandy. Earlier, when the North Dike was being built, engineers had avoided dense top soil, laden with seeds, roots, and at times ice, because they were concerned that it would weaken the embankment. Instead, based in part on their testing program, they laid down sand.

This choice had a serious consequence. On April 11, 1907, as water slowly filled the reservoir to full capacity, about 46,500 cubic meters of soil, from a 213 meter section of the dike, gave way sliding *into* the reservoir. The soil came to rest below the water, but it did not significantly affect the capacity of the reservoir. More soil was used to fill the liquefied portion, and the dike was repaired by October 30, 1907 with engineers choosing a lower slope angle and in some places adding more riprap. Thankfully, the failure had been relatively minor, and the entire dike avoided liquefaction and flooding.

And what of the thirsty city for which these tasks had been undertaken? In fact, the task of transporting water back to Boston from the first aqueduct work begun in 1896 had been on a monumental scale. Two miles of the new aqueduct had been cut through rock, another seven miles at or near grade and lined with masonry, including the handsome arched aqueduct that spanned the Assabet River, in Northborough.

By 1905, and even more by 1906, with construction wrapping up, the shantytowns of the workers were being dismantled and cleared away. The workers that hadn't put down roots in the community moved on to find work in other towns, or other states. Some of the skilled laborers and engineers left to work on the dam in Katonah, New York, as part of New York City's expanding water supply system.

Another indicator of the project's state of completion was a reduction in on-site policing, summarized in the 1905 Water Board report. "For the Wachusett Reservoir district there have been employed 18 officers: eight (reduced to three in December) in the town of Clinton; three, one of whom is mounted, (reduced to one in December) in the town of Boylston; six (reduced to five in November and placed on half time in December) in the town of West Boylston; and one mounted officer in Sterling."

In the intervening decade, between the start of construction, and the completion of the dam, locals slowly became used to the large immigrant presence (although they often still resented it) and had come to tolerate the monumental changes happening in their 'backyard.'

After completion of the dam, and notwithstanding the liquefaction of part of the dike in 1907, the reservoir continued to fill as the Nashua and Quinapoxet poured their water into the vast basin, and regular rainfall stepped in to help out. The reservoir was finally declared full in 1908.

That year was both a happy and a melancholy one for West Boylston. The town celebrated its centennial. One hundred years earlier, it had broken off from Boylston. Now, amidst the festivities, West Boylstonians were reminded of how much their town had lost, and how it had changed.

Clinton was excited too. The town government was in the process of planning the new town hall, the same one that stands in the center of town today. The reservoir had changed the way in which Clinton looked at itself. The town was now home to one of the largest dams in the world, adding to its list of technological achievements. At the same time, the town had been forced to give up land, and watch as the great stone barrier was built across its primordial thoroughfare -- the Nashua River.

Although the Water Board made much of this achievement, for the valley towns that had watched the day-by-day progress for years, completion of the reservoir seemed to be almost anticlimactic.

Indeed, sources of news and excitement as recorded in the *Item* through 1908 often derived from other aspects of the community. For instance, a rash of hydrophobia swept the Clinton during the winter, and as spring rolled around, several people were bitten by rabid dogs. However, the *Item* found a hero in one Gordon Crossman, who took the matter, quite literally, into his own hands:

> *Mad Dog Killed-Gordon Crossman Grabbed the Animal by the Neck and Holds Him—Until Axe is Brought*
>
> *A German shepherd, afflicted with rabies roamed Clinton, biting other dogs. A local man spotted the dog, and proceeded to chase it across town. Following the dog for possibly several miles, Crossman grabbed it by the neck, and held it at arms' length, as he retraced his steps in search of an axe. When he finally secured an axe, he promptly bludgeoned the mad dog to death.* (Daily Item).

After Gordon Crossman's exploit, the other dogs that had been nipped by the rabid German shepherd were killed, helping to end the menace to public health.

At the beginning of the year, Walter O'Malley, a fifty year old local man, fell through the ice of Mossy Pond while fishing. Later in the year the *Daily Item* nonchalantly reported the death of unidentified Eastern European man. His effects had been found on a bush nearby, and there were letters written in Lithuanian or Russian, in his pockets. No one seemed to suspect foul play, but the death was never properly explained. A more humorous note was sounded by the *Daily Item*, when the newspaper reported on July 22, 1908, a skinny dipping incident at the Lancaster Mill Pond.

> *Swimming at the Bridge—Complaint made to the police about nudity of bathers in Lancaster mills pond—Repeated instances*

> *The past four or five days, at about 5 o'clock there have been seen within a short distance of the Lancaster mills bridge young boys of varying ages in swimming and they were as naked as when born.* (Daily Item).

Other, sensational local stories included the occurrence of food poisoning at Lancaster Academy (the corn in the noodle soup was to blame); the Knights of Pythias convention in Boston; the discovery of an intact 18th century cannon in the woods by a local boy; the arrival of a socialist from Texas; a selectman's son injured by a cannon during a holiday celebration; the opening of a tuberculosis day camp; and a story about an elderly Clintonian, James J. Walsh, supposedly the last man to convey a message to General Custer before his demise at the Little Big Horn -- all found their way on to the front page, seemingly generating more attention than the reservoir. Ironically, many Clinton townspeople were unable to take advantage of the immense reservoir for recreation. Unlike the many small ponds that dot the Massachusetts landscape, no launches or steam boats were allowed on the reservoir, unless they were on state business, and furthermore, swimming was prohibited. So Clintonians, not satisfied with the remaining local swimming holes, could look forward to a few days on the shore of Lake Winnipesaukee, if they could pay the $1.85, round trip fare, to the Boston & Maine Railroad.

Meanwhile, a drought raged throughout the summer and early autumn of 1908, raising concerns that the reservoir's water level would drop too low and an *Item* story hinted at consequent agricultural distress:

> *A Boston daily of today's date speaking of the possibility of a shortage of the milk supply says: 'just as serious is the possibility of a water shortage in the metropolitan water district which has not been deemed possible since the big Clinton reservoir was built.* (Daily Item).

The author of the story expressed confidence in the ability of the reservoir to sustain the water loss (the water dropped as much as an inch a day), pointing out that the Wekepeke Brook and aquifer (in Sterling and South Lancaster), were still safe from evaporation. Some relief from the drought came in a very bizarre weather incident. On July 16, 1908, after weeks of hot dry weather, with temperatures of 94 degrees Fahrenheit or higher, a sudden snowfall coated the ground in as much as an inch of snow. Snow in July is an unusual under any circumstances, but the fact that so much fell is truly an indication of the oddities of New England weather, and Earth's climate (though one might suspect that this event was related in some way to the famous Tunguska incident in Siberia on June 30).

Whereas in Clinton, the promenade and fountain below the dam were open to the public, on the western end of the reservoir, there were only the utilitarian aspects of the project and the Old Stone Church to remind people of old times. Still, the image of the reservoir at the time of its opening is of a vibrant and exciting place – a popular subject for purveyors of penny postcards. People from Clinton, the surrounding towns, and the wider world arrived to see the new marvel. The tumultuous years of the reservoir's construction, the sweat expended, and the lives lost, were fading into the past.

Clearing the Land

Thousands of trees were cut to make way for the rising waters. Workers stacked hundreds of cords in neat piles such as these. (Clinton Historical Society)

Seen in the background of this photo of Dumping Platform No. 20 in West Boylston, taken October 7, 1901 are clumps of burning brush. (Clinton Historical Society)

Taken from the remains of Holbrook Mill, this photograph shows Cowee's Grist Mill across the Nashua River. Note the abundance of tree stumps in the foreground. (C/W Mars Digital Treasures)

A man hitches a ride on a soil cart, making a stop at a dumping platform. Three horses and their accompanying carts are queued up to unload. (Mass. Archives)

A close up of one of the many soil carts used in the Wachusett project. It appears that the wooden structure in the upper right hand corner, may be one of the dumping platforms. (Clinton Historical Society)

Wagons pour soil down the triple chutes of this impressive dumping platform. (Clinton Historical Society)

A group of workers poses proudly with their pneumatic drills, carts, and donkeys. (Clinton Historical Society)

Cunningham Grove was a popular get-away in South Clinton, hosting parties and banquets. Mr. Cunningham appears on the left hand side of the photograph. Cunningham was one of the last property owners in Clinton to sell out. (Clinton Historical Society)

Wachusett Dam

Excavation is carried out for the Wachusett Dam, in June 1903. The wooden structure at left was used to divert the Nashua River around the excavations. (Clinton Historical Society)

Workers armed with high-tech rock drills chip away at the seemingly impenetrable bedrock where the planned dam is set to take shape. (Clinton Historical Society)

The first stones of the dam show above the ground, in this winter photograph. Note the barrenness of the surrounding land. (Clinton Historical Society)

Boilers emit clouds of steam as the foundation of the dam begins to take shape. (Clinton Historical Society)

Four 48-inch pipes laid down to carry water through the dam's foundation. (Clinton Historical Society)

Two workers, each smoking a pipe, pose for the photographer beside this 48-inch pipe, on September 6, 1902. (Clinton Historical Society)

The nearby quarry that supplied stone for the Wachusett Dam. The ashlar stones on the outside of the dam were quarried in Chelmsford, this pit provided the interior stones. (Clinton Historical Society)

The very first stone in the Wachusett Dam was an early milestone in the dam's constructions. Workers celebrated by breaking off chunks of the stone, with hammers, as souvenirs. (Clinton Historical Society)

A view across the valley from Burditt Hill, showing the four cableway towers erected for the project. The cableway was used to carry slabs of stone to different points on the dam, where they were cemented in place. (Clinton Historical Society)

A horse and carriage approach the cableway towers. (Clinton Historical Society)

Like gothic monsters, twin cableway towers crown the hillside. There is no cable shown in the photograph, so the towers may have been newly erected. (Clinton Historical Society)

An up-close photograph of one of the cableway towers. (Clinton Historical Society)

A man cutting slabs of stones peers at the photographer through goggles. Many skilled laborers later packed their bags when the project was finished and left to work on New York's reservoirs. (Clinton Historical Society)

Workers pose in the cableway basket used to carry stone slabs across the valley. (Clinton Historical Society)

The fountain nears completion. The horse must have hoped that that the fountain was an enormous watering trough. (Clinton Historical Society)

Construction on the Wachusett Dam and power house speeds along (Clinton Historical Society)

A view of the Nashua river valley, the completed viaduct, and the still infantile dam, from the present location of the Museum of Russian Icons. (Clinton Historical Society)

The pump house takes shape as the cableways carry stone slabs to the three platforms to be inserted. (Clinton Historical Society)

Water and construction debris piles up behind the partly completed dam. (Clinton Historical Society)

A spectacular photograph taken from the viaduct, shows the derricks and cranes hard at work on the dam. (Library of Congress)

A view of the reservoir face of the dam; note the box cars in the left foreground of the photograph. (Clinton Historical Society)

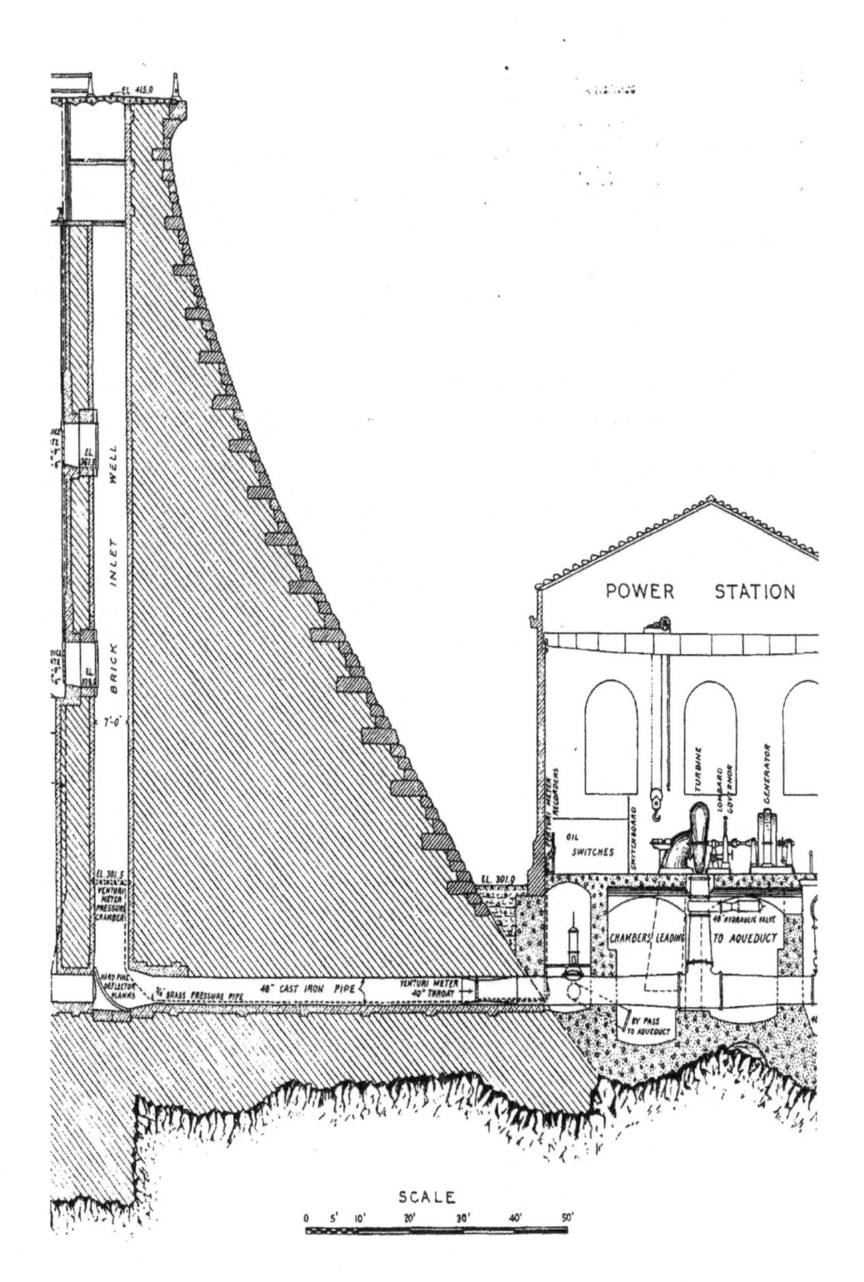

This excellent cut-away shows the network of pipes, chambers and meters that make up the bowels of the Wachusett dam.

Laying of the last stone on the Wachusett Dam, June 24, 1905 (Grady Family)

New Routes

Taken at the west portal of the C.M.R.R. tunnel, on November 11, 1902, this image gives some idea of the size of the tunnel. The tunnel was the second longest in Massachusetts after the Hoosac Tunnel, in Florida, Massachusetts. (Clinton Historical Society)

A team of horses pull what appears to be a water tank, followed closely by a steamroller. (Mass. Archives)

The aqueduct crossing the Assabet River in Northborough is an impressive piece of architecture and one of the first engineering successes of the project. (Author's Photograph)

Erecting the Viaduct

The early stages of the viaduct's construction appear surprisingly similar to the area today, where the piers still stand. Note the beams in the foreground, and the entrance to tunnel on the hillside. (Clinton Historical Society)

A passenger train steams across the viaduct. A small crowd watches from beside the fence. (Clinton Historical Society)

A train crosses the viaduct in front of the completed dam (Clinton Historical Society)

The Spillway

This bridge, that still stands today, is seen from the spillway gates. (Clinton Historical Society)

The Wachusett Reservoir was acclaimed as a technological marvel, and every aspect of the dam was recorded in souvenir postcards- even the spillway. (Author's Collection)

A good overhead image of the lower spillway bridge, where excess water runs into Lancaster Mill Pond, soon after its completion. Some horse drawn carts are already making use of the bridge. (Clinton Historical Society)

North Dike

The North Dike is a hive of activity on July 5, 1904. In the foreground are a gang of workers as well as horse drawn carts, while in the background the narrow gauge railway cuts through the picture hauling soil. (Clinton Historical Society)

A rare view of the cut-off trench, and what was essentially an underground dam designed to stop seepage that would undermine the North Dike. (Clinton Historical Society)

The North Dike nears completion on February 6, 1903. Today, the platform at the left would only be a few feet above the water line. (Clinton Historical Society)

The Reservoir Fills

Between 1906 and 1908, construction had largely ceased, as the reservoir began to fill. (Clinton Historical Society)

Another view of the reservoir as it filled. The hills in the background are strikingly barren. (Clinton Historical Society)

As the reservoir filled to capacity, West Boylston celebrated a bittersweet centennial as shown here (C/W MARS Digital Treasures)

Work and the Workers

Rome wasn't built in a day, and neither was the Wachusett Reservoir. When the project began, the water board realized that it would need to hire many laborers -- thousands. Most Clintonians, and their fellows in neighboring towns, were already gainfully employed by the Bigelows and other mill owners. Although there had been some cuts in hours because of the economic recession, most people were employed, and few people were willing to take on the low-pay and grueling physical labor on the reservoir project. The project required hundreds of laborers and at a time when factories were probably not hiring as much (due to the economic recession of the early 1890s) the Wachusett project offered steady employment. In fact, because of the national economic slump, companies such as the Lancaster Mill, cut their schedules, so that employees were only working half the total number of hours that they had worked before. Thus, the project, and the secondary business it brought to the area, probably helped tide over some Clintonians; it certainly helped landlords, grocers, and others who suddenly had many new customers.

Who were the "new" people that came to work on the project? Most were Italian, Hungarian, or Finnish. There were others too, including a substantial group of African Americans, willing to put up with cold winters in exchange for pay, who emigrated from Virginia. What was true of all, was that they were willing to work, and work they did. Laborers' sweat and dedication saw to the clearing of the reservoir's bed, the building of the dam, the clearing of trees and houses, and all of the other tasks. All told, 4,000 overseas immigrants and laborers from the Southern United States arrived in the region. Life was certainly hard for the laborers, and at times dangerous, such as in the case of the Michael Holloran – the project's first casualty. A total of 37 people died in the building of the reservoir. The highest number of fatalities occurred in 1897, with nine people killed. Cave-ins and explosions were at fault for some deaths, but workers were also crushed between pipes, killed in falls or struck by trains.

The 1905 Water Board report chronicled five fatalities for the year: "…three at the Wachusett Dam, one upon the railroad from the quarry to the dam, and one upon the Wachusett Reservoir."

"The first four accidents occurred in connection with the work of the McArthur Brothers Company. At the dam a masonry inspector had the bone of his thigh broken by being caught between a stone which was being raised and the masonry of the dam, and as a result of his injury died thirteen weeks later; a masonry foreman was killed by the falling of a boom of the derrick; and a laborer by the falling of a large stone which slipped from the grab-hooks. A blacksmith was killed by a collision on the quarry railroad. On Section 8 of the Wachusett Reservoir a laborer was killed by falling under a train on the contractor's railroad."

Notwithstanding such grim statistics, (and it is worth noting for comparison that coal mining accidents sometimes killed thousands of Americans a year in this era), for new immigrants, who had arrived in search of work, the Wachusett project was apparently a godsend. As noted above, while a handful of local workers found employment on the project, the effort was dominated by immigrant labor. Only about one in every 20 workers, had been born in the United States.

Underscoring the reasons why native-born workers were so rare, an 1896 article about the reservoir project, in the *Worcester Telegram* noted, "…few are willing to bend over pick and shovel under the rays of a scorching sun all day long and receive for this work the small pay which the contractors…are giving laborers. Therefore, the clause in the Metropolitan Water Act which provides that citizens of the commonwealth shall be given preference amounts to nothing."

Indeed, one could hardly expect to get rich off of the wages paid by the project. Unskilled laborers, who had only their muscle to offer, were paid between a $1.25 and $1.50 per day. If you were a skilled laborer -- a stonemason, a carpenter, or a blacksmith -- you could hope to earn from two to six dollars a day.

The contractors on the project worked relentlessly to keep up progress. To accomplish as much as possible, they sometimes fed on interracial or interethnic feuds, blowing them out of proportion, so that each ethnic group would compete harder and pull ahead of whatever race or group it was in conflict with. The *Daily Item* described instances of violence between the races, as the, "Pent-up fires of color and racial jealousy." This led to at least one instance of violence. A black laborer crushed the skull of an Italian worker, mortally wounding the Italian.

Conflicts arose between the settled townspeople, the Water Board, and the laborers, from time to time. Many of the issues were labor disputes, while others involved alcohol. Ironically, it was many of these immigrants, often shunned by locals, who put down roots, and whose descendents live today in the reservoir towns. If it had not been for the reservoir, these people and their families – so crucial to the region today -- might never have found their way to Clinton, Boylston, West Boylston or Sterling.

Aside from the competition that the newcomers provided for jobs on the project, the influx of so many immigrants was also sudden and unprecedented in its scale for the region. Some 3,500 new workers were present in the area by 1897. Where to put them and the extent to which they would be welcomed in the community as shoppers, congregants, or citizens was an open question.

Jim Frank, an Italian who would experience a run in with the law before the project was completed, had tried as early as 1895 to rent a building to house one hundred workers, for whom he would supply food and board. The local real estate board decided that Frank's actions were premature, and dissuaded him, but Frank eventually got his wish, and rented a building.

Even when they earned low wages, and worked in poor conditions, American life was still apparently regarded as much better than in "the old country." The thousands of Italians, who had immigrated to work on the reservoir project, could hope to earn as much as six times what they could in Italy. This was true even for *braccianti*, unskilled laborers. The better wages across the ocean in America attracted many Italians, through the so-called 'padrone system.' A *padroni* was a fellow Italian such as Frank (sometimes a family relative), who had immigrated to the US earlier, and had acquired knowledge of English and the workings of the local job market. When new immigrants arrived, the padrones would help to find them jobs. Because most of the new immigrants could not speak English, they were often hired to do what Italians called *sciabola* labor -- work with a shovel. Due to the discipline provided by the padrone system – workers owed an obligation to the padrones -- Italians labor became the driving force behind much of the reservoir project.

Although at first glance, the aid and guidance provided by a *padroni* seems as if it would have been beneficial for the new immigrants and the *padroni,* unfortunately, the padrone system often turned out to be a one-sided bargain; good for burying new immigrants under a mountain of debt. For example, the padrones often had the sole right to house, feed, and hire out the new immigrants at work on the project. Even if a laborer did not live in one of the padrone's shanties, he still had to pay a dollar a week to the padrone. The padrone often received all of the workers' wages, and then paid it back to them, after detracting the five dollar charge demanded for helping the new immigrant find work. But this "take" often extended further, with the padrone's charging double, or even triple the cost in town, for low quality food. Pork, which cost at most six cents a pound in town, could cost 18 cents a pound when buying from a padrone.

Out of the dissatisfactions of this system, labor disputes arose often, as laborers stood up to powerful padrones and contractors, going on strike, or forming a temporary labor union to demand better working conditions. This was not always a good course of action, because workers could be dismissed by a padrone, and new arrivals hired in their place with little difficulty.

One instance of the unfair working conditions was reported on March 11, 1904, by the *Daily Item.* Italian laborers protested the housing agreement under which the Mcgee Company operated:

> *J.P. Mcgee and Company are facing another period of trouble with their Italian employees on the South Dike contract, and this morning's prelude to events portends some serious happenings before the difficulty is all over. Two weeks ago today Italians stopped work throughout the job. Today there was a complete suspension, but conditions are different. Last time the men acted practically as a unit. There were very few when the strike was inaugurated, who desired to continue at work.* (Daily Item).

The company provided shanties for laborers, and prevented most laborers from living elsewhere. Laborers with families were angry, because they did not want to live in inadequate shanties provided by the company. Trouble resulted when a foreman tried to fire non-shanty dwellers, leading to blows. In this case, the Clinton police were telephoned, but the disturbance had been quelled by the time they arrived.

Shanties, the catch-all term for the workers' makeshift dwellings, varied from small shacks made of cast off wood and debris to lean-tos, thatched with pine boughs. Some were hardly weather proof at all and offered only the most rudimentary shelter.

Although the Italians and other white workers suffered from discrimination and abusive practices, and locals seemed to hold their shantytowns in low regard, it seems the most underappreciated workers on the project, were the black laborers, many of whom had emigrated from the Southern US, where labor conditions were deplorable and accompanied by the repression of the Jim Crow period. Even in the more racially tolerant North, the black workers' lot was poor. They were given some of the most dangerous jobs, such as planting dynamite and excavating underground for the aqueduct (the Water Board annual report noted blandly that 300 workers, "mostly negroes" were employed on drilling the rock tunnels). Not surprisingly, black laborers suffered some of the highest numbers of injuries and deaths out of all the different ethnic groups represented on the project.

The black shantytown, known as Smithville, located near Shaft 1, was seen by Clintonians as a hive of villainy – and occasionally even murder. In 1897, the *Worcester Telegram* followed a Clinton court case, involving black laborer John Christmas, who had a running feud with Caroline Johnson, his lover and caretaker of his shack. It seemed that Caroline Johnson had a love affair with another man, leading to a fight with Christmas. Christmas was fined ten dollars by the court. In another instance, when Clinton had workers on the dam project arrested for failing to pay the local poll tax—an action that smacks of harassment -- most were quickly freed when friends or their employers came up with the tax money. However, one forlorn African American was left to languish in the Fitchburg jail for an extended period of time awaiting disposition of his case.

Somewhat more benignly, black workers were singled out for a special "honor" when Clinton staged a cakewalk that was attended by one hundred black laborers and their female friends or wives. The cakewalk was a form of minstrel-like dance from the South, left over from the days of slavery. Eleven couples at a time, performed before five white judges, while a white audience from town looked on. The first winner was Edward "Happy Ned" Nash, employed on the aqueduct portion of the project, who, along with his dancing partner, Mary Page, was awarded a five foot tall cake.

Beyond such special occasions, the relations between townspeople and the project workers remained tense. Clintonians, for example, did not seem to mind Hungarians and Irishmen drinking in the town saloons and bars, but they were alarmed at the arrival of Italians whom they seemed to view more negatively. Thus, the town encouraged Italians to buy any liquor they might drink from a commissary. The plan of thereby supervising the Italians' drinking habits, and keeping them away from town saloons, backfired. The padrones circumvented the existing liquor laws by selling alcohol under the counter, and failing to record each sale. Padrones were able to get away with such violations, perhaps because the Metropolitan Water Board had no objection to laborers drinking away their money, when they were not working on the project.

Because of the hard conditions, and the frequent stopping and starting of work, many immigrants turned to alcohol to overcome everyday hardships, while padrones and other immigrants became suppliers of alcohol to augment their limited income. The alcohol problem eventually led to raids, searches, and arrests by the Clinton Police, and the Metropolitan Water Board.

Alcohol and housing became perennial sources of conflict between the workers and existing communities and among the workers themselves. On March 17, 1904, two Italians fought each other with a bottle and a pistol in a tenement in Clinton. No one was injured, but one of the men was arrested. On another date, two laborers butted heads over the issue of which one of them was the rightful owner of a "grocery shop," set up in a shack at the North Dike. Problems like these became run of the mill disturbances for the Clinton Police.

Valley communities also had to adapt to the fact that they were now dependent on a state agency. For instance, along with providing its own small police force, the Metropolitan Water Board provided for local policing of the project by paying the salaries of (at one point) nine out of the 15 Clinton police officers. Clintonians did not like the fact, that, in effect, even their police department was now controlled from Boston, and that the Metropolitan police interfered with their jurisdiction. This process was described in a Water Board annual report as follows:

"Police protection has been afforded, in according with the requirements of the Metropolitan Water Act, in places where active construction has been carried on. The police officers have been appointed by the various towns in which their services have been required, and they have been subject to the town authorities in the performance of their duties, but they have been paid for their services by the Board."

Another bone of contention was the new patrol wagon that the *Item* reported about on July 18, 1899.

> *The Patrol Wagon*
>
> *The Metropolitan Water Board recently granted to the town of Clinton a patrol wagon and three metropolitan policemen. The wagon was believed to be as fully essential as any of the three officers and was to greatly facilitate the bringing in of drunks and disturbers who have their existence in Clinton largely on account of reservoir work. Until the arrangements which the Water Board had made became known to the board of selectmen and the police department it was thought that next to having the six officers which were desired, no arrangement could be better than that self-same hurry up wagon.* (Daily Item).

However, much to the irritation of the Clinton police department, the wagon, because it was privately owned, could only be used by Metropolitan police to make arrests. "The other officers of the local force are compelled to resort to the old practice of pressing into service a hack or a grocery wagon or any other vehicle which may be handiest at the time." Under that older arrangement, 50 cents per arrest were to be paid to the livery men and the Clinton police simply had to make do.

Later in the month of July 1899, on the 24th, the *Item* reported another of the frequent liquor raids, which brought a sizable crowd of arrestees.

> *The first raid was at the place of Jim Frank, who has been before the court before for illegal liquor selling. His shack is on the place of the temporary road not far from the old St. Johns cemetery...Frank's place has accommodated as many as 100 men, but there are only 50 there at the present time.* (Daily Item).

The padrone, Jim Frank, had been making additional money (after food and board) off of his laborers, by selling them liquor.

Of course, the shantytowns weren't the only place with problems getting quality food. On July 8, 1901, a Clinton man was arrested for selling stale fish -- it was mackerel -- that was decomposing. Presumably the offensive smell was sufficient grounds for the arrest!

The Laborers

Workers of Italian and Hungarian extraction, and the contractors who hired them, built shantytowns that lasted for the duration of the project. Squalid settlements such as this clustered in areas such as the North Dike, and were often divided by ethnicity. (Clinton Historical Society)

Shacks such as these were assembled from lose boards and pine boughs. (Clinton Historical Society)

A group of workers pose proudly around a railroad crane owned by one of the contractors for the reservoir. (Grady Family)

Labor Trouble

Almost from the start of the project, labor trouble was an accompaniment to the project. Researcher Jill Lepore noted that the *Clinton Daily Item*, the longest running paper in Clinton, tended to take a critical view of the laborers. The opposite was true of the *Clinton Courant*, which was more of a working man's paper. Many stories relating to the labor disputes were hardly covered by the *Daily Item*, and in fact gained more attention in the pages of the *Worcester Telegram*.

As early as 1896, Metropolitan police officers had been called on to quell a strike. Workers were furious because they claimed they had not received any pay in at least three months. All but one out of two hundred workers went on strike. The strikers were kept under control for three days, with promises that their pay would arrive soon. However, this was not to be. The contractor had skipped town with the payroll!

Even though most of the laborers could hardly speak English, many of them found their way into town to complain to public officials. Orrin Bates, the Clinton chief of police, frequently met laborers who came into town in small groups to file complaints. The Chief's response was always, "Of course it is out of our line entirely." The Metropolitan police were equally unsympathetic, and ignored the majority of the complaints brought to it. A further complication for Italian laborers was the rural location of the reservoir project. In the cities, Italian-American clubs and unions could help out impoverished workers, but here, in northern Worcester County, the padrones and contractors seemed to hold all the cards.

However, the plight of labor did not go entirely unnoticed. Throughout the course of the project, the laborers were able to make some allies in high places, who helped to remind others of the workers' plight. One of the most significant of these allies, was writer Lynn Wilson, a reporter for the *Worcester Daily Spy*, *Telegram*, and the *Boston Globe*. Wilson's articles were sympathetic to the workers. The Clinton Businessmen's Association also took the side of laborer's at times. In particular, the CBA opposed the padrone system. A local pharmacist, Thomas Tate, wrote up a report on the padrone system that was published in *Worcester Daily Spy* in 1899. Wilson also spoke out in favor of reelecting David I. Walsh as state representative, because of his anti-reservoir, pro-labor sentiments.

Countering the pro-labor forces of Rep. Walsh, Lynn Wilson, and the Clinton Businessmen's Association was the Water Board's lawyer J. Benton, Jr., who collected information to prove that Clintonians were in support of the reservoir, and that no Clintonians had ever thought of working on the project. Benton collected statements from businessmen not associated with the CBA, many of whom profited from selling supplies to the Water Board, to the effect that they had never heard any complaints made against the Water Board. Moreover, Benton accused the CBA of taking an interest in the workers only because of its own business interests. This was partially true, for the CBA was primarily composed of the businesses in Clinton that were in part fed by laborers' wages.

Thus, Benton seems to have had success in portraying the CBA as a less than disinterested group of businessmen, in fact out to make a profit without having true concern for the worker's interests.

Meanwhile, on February 27, 1900, the state legislature had called a special committee to investigate the Metropolitan Water Board and the accusations made against it relating to labor conditions. The Water Board had, among other things, purportedly failed to show preference to regional laborers and favored the padrone system. However, the committee investigation bordered on farce. It was held in Boston, making it difficult, if not impossible for poverty stricken Italian laborers to attend and testify (only three laborers testified). Out of 91 witnesses, 27 were members of the Clinton Businessmen's Association. An equal number were from the Water Board. At the start of proceedings, Representative Walsh asked that the committee meet in Clinton, but the request was promptly denied.

The CBA responded by calling a meeting in Clinton on March 30, 1900, to discuss working conditions. The meeting began with a brass band parading around town. By early evening, the town hall was packed with 1,200 or more residents. The meeting was 'impartial' but everyone was apparently in agreement that the Water Board was in the wrong.

Days later, a smaller crowd of socialists from Clinton met to talk about the meeting. The socialists were determined to consign the padrone system to the past, while also aiming to double wages for the laborers. They believed that the CBA was not forceful enough and would get workers nowhere.

As the special committee investigation played out, Henry Sprague, the head of the Metropolitan Water Board, insisted that he had no knowledge of the padrone system. Sprague probably wasn't lying since the heads of the Water Board seldom made their way out to Wachusett from Boston.

As justification for hiring immigrant laborers, the Water Board stated that the cost of hiring local workers in the first five years of construction would have amounted to an additional million dollars. The committee issued its report in July of 1900, refuting some of the accusations against the Water Board pertaining to the treatment of local labor. "If aliens are willing to do as good work more cheaply, the duty of preference does not require the employment of citizens at a higher rate of wages." However, the committee did state that the Metropolitan police had failed to enforce labor laws, and recognized the issue of the padrone system, which it condemned. As a result of the report, a new law was passed, that effectively outlawed the padrone system.

With the padrone system a thing of the past, worker's conditions improved as the 20th century began. In fact, the Clinton Businessman's Association and its support of the Italian laborers was one of the earliest examples of anti-padrone activity in the United States. Nationwide, organizations such as the Society for the Protection of Italian Immigrants continued to work to end the unfair system elsewhere.

In a letter to the editor of the New York Times, Richard W. Gilder noted:

> *The great increase of Italian immigration necessitated the formation of this society, and in its two years of existence, under the persevering and thoroughly businesslike Presidency of the young lawyer, Mr. Eliot Norton, it has accomplished a great and vastly needed work, distributing to immigrants without charge, thousands of dollars sent by the friends of the newly arrived; informing, advising, and protecting numberless bewildered Italians; rendering valuable assistance to the United States Immigration authorities; escorting thousands of friendless immigrants, at low charges, to their destination, and finding employment for hundreds of laborers.* (Daily Item).

The committee report did not mark the end of strikes on the Wachusett project. These continued from time to time, such as in 1904, when Italians protested shanty rents of 25 cents a week. On the whole, though, the workers were better off than before the investigations. The contractor Nawn & Brock even increased workers' pay, because they were short handed.

The workers' struggle for better living conditions, better wages, and an end to the padrone system was a backdrop to the Wachusett project. In a very real sense, the reservoir was built and paid for with the workers' sacrifices and hard work. The immigrant laborers did not always receive good treatment from Clintonians, the Water Board, or the residents of the other reservoir towns. However, in some cases they were able to come to agreements with local businessmen that proved beneficial for them, and secured wage hikes by standing up to the Water Board, the contractors and the padrones.

A contemporary, sympathetic to the workers, penned a poem that aimed to commemorate the reservoir workers and their sacrifices:

The Building of a Waterway

The Work on the Metropolitan Dam, Clinton, Massachusetts

We stood on the brink of a cavern,
And gazed in the depths below,
The gloom of the night was round us,
Lit up by the torch's faint glow.

Fair down in the darkness beneath us
Men bent o'er their wearisome task:
Like dwarfs, or like pygmies, they labored,
Half hid by the night's gloomy mask.

The water oozed darkly around them,
Above them, the great engine roared,
Showing bright, by the glow of its furnace
Where its mighty life forces were stored.

The work looked so slow and so hopeless,
No trace of design could be seen;
Just groping and toiling in darkness,
Without plan, or purpose, or scheme.

Thus it seemed, 'till I lifted mine eyelids,
To the country, that stretched far away,
With its hills, and its woodlands, and valleys,
That will vanish, like phantoms, some day.

For I saw, with a clarified vision,
The glory and wonder to be,
The path of a might river,
To the city down by the sea.

Over acres, and acres of country,
The shimmering waters spread,
Bringing comfort, and health, and refreshment,
In its glittering wake, as it sped.

Health for the toiling millions,
In years that are yet to appear;
Comfort to thirsting people,
On its broad banks far and near.

And the valleys shall be exalted,
And the mountains shall be laid low,
To carry the flood of waters
Where the city waits below.

But the toilers, there in the darkness,
Knew little of purpose of plan,
So great, that it seems like a fable
Or a miracle, wrought by man.

Yet each had his task, or his portion.
And each was a part of the whole;
And naught could be shifted nor slighted,
That might keep the great flood in control.

Then there came to my mind, as I pondered,
A glimpse of creation's great soul,
Where each is a part of the other,
And all are a part of the whole.

Our work may seem petty and toilsome,
We may strive in the darkness alone;
We may sigh, at the mountain before us,
And weep over failures outgrown.

Yet, lifting our eyes from this earth task,
And seeking to follow the light,
We may rise from the dark slums of matter,
And pierce through gloom of night.

With the light of the knowledge before us
That we are a part of other plan,
We can work and trust to the Father
His infinite purpose for man.

---Carrie Cutting Adams, 1900 (Clinton Historical Society)

Today, after more than a century, Carrie Cutting Adam's vision of progress has been fulfilled, with the Wachusett as a critical piece in a water system that has grown even more substantial. Yet, the full recognition of the workers remains a task for the future. Unfortunately, there is not even a monument or plaque commemorating those who lost their lives or suffered injuries bringing the project to completion.

Changing Communities, Changing Lives

The building of the Wachusett reservoir brought great changes to the towns which surrounded it. For one thing, Clintonians could no longer reach Boylston by their familiar routes, and vice versa. Not only was navigation altered, the landscape was often changed beyond recognition. (Though town border lines remain as they were, beneath the waters of the reservoir.) It is hard to picture exactly what a West Boylstonian, or anyone else passing through the western end of the reservoir, would have seen. Dozens of houses, abandoned, and going to ruin would have the most noticeable feature. There would probably have been many indications of the ongoing construction, such as crowds of laborers, tree stumps, and wood piles.

In fact, parts of Oakdale and West Boylston were transformed into ghost towns, and the reservoir's bed at its western end was littered with abandoned properties. A traveler setting out on foot or on horseback from West Boylston center, bound for Clinton or Lancaster, would first pass by the house of C.H. Baldwin. The original Baldwin house had burned to the ground 50 years before, and had been replaced by a smaller house. But although this familiar landmark escaped the wreckers, it too was missing in action; thanks to its small size, the Baldwin house was moved, and survived the building of the reservoir.

In fact, most West Boylstonians had sold their homes in the reservoir bed in 1896 or 1897, when the State offered a good price for the houses. Not only houses, but businesses were bought up. Several centuries of West Boylston history were swept clear within a few years. Local employees of bought up companies, were forced to either move, or find new employment. Many found work at an organ shop owned by George Reed, but Reed could only so hire many workers. Starting in 1898, certain houses (including the Baldwin house) were removed from the reservoir bed, and relocated elsewhere in the town. A town meeting decided that houses could be relocated to the common, which spawned another dispute over the future of the trees on the common.

By 1899, most of the remaining 38 houses targeted by the water project had been demolished. The good wood from the houses was sold to build new houses, while rotten wood or wood of lesser quality was sold as firewood. Windows and doors were salvaged and resold.

Where the town had boasted 484 dwellings in 1895, a decade later there were a mere 250. Likewise, between 1895, and the sorrowful town centennial in 1908, the town's population was halved from 2,968 to only about 1500. By then the reservoir had filled, and all that remained of a large swath of West Boylston and Oakdale was memories, and a few material reminders at the bottom of the reservoir.

Among the local icons sacrificed to build the reservoir was the Beaman Oak, a part of local lore even in the 21st century. By some measure, the largest white oak tree in Massachusetts, legend had it that Ezra Beaman had planted the Beaman Oak, when he was a young man by sticking his oak branch riding crop into the ground. Since Ezra's descendants had prospered, the family name was well represented among the casualties in West Boylston: The Beaman Cemetery was uprooted and the Beaman Tavern was demolished. Thankfully, however, the Beaman watering trough was donated to the town as a keepsake and made it to higher ground.

The Beaman Cemetery had a place that was nearly as special in town folklore as the Beaman Oak. Oddly, for the 60 or so corpses buried in the cemetery, only one casket was found, presumably belonging to Jabez Beaman, who died of smallpox in the 1750s. In order to move the casket, all of the Beaman descendents had to sign their approval for moving the cemetery plot and their ancestor's body. One house, built in the 1790s, and one of the oldest in the town was also torn down. A 25 foot tall, stone chimney remained for some years after the house was gone.

Some of the houses demolished in West Boylston unwittingly revealed hidden secrets. For instance, several houses owned by well known 'teetotallers' turned out to have secret trap doors and compartments for storing liquor, some very elaborate. Ironically, this construction that so disrupted day-to-day life, proved to be a solution to an unforeseen problem. In the summer of 1901, a smallpox outbreak terrified West Boylston. Laborers were hurriedly vaccinated, but the handful of people who had caught the disease needed to be isolated to prevent its spread. To this end, one of the remaining abandoned houses in the reservoir bed was used to quarantine victims. Each day, one local gentleman, who had previously survived a smallpox infection, would come down to the house, to pick up a grocery list left on the front step by the sickened inhabitants.

By contrast with West Boylston, Boylston and Sterling were comparatively unscathed but the effects of the project were still significant. Boylston and Sterling were favored by geography in that less of their population and buildings were along the Nashua's valley. It should not be forgotten, however, that Boylston lost the manufacturing village of Sawyer's Mill, as well as Albertson's Mill. Although the Nashua was not actually their boundary, its enlargement into a reservoir only enhanced a natural division that was already present between the two towns.

For Clinton, the building of the reservoir meant dramatic changes, too; the dam, a new railroad connection, and an improvement in the town's sewerage system. Until 1898, Clinton continued to pour raw sewage into the Nashua River. We may grit our teeth, and turn up our noses at the thought, but disposing of sewage in this way -- dumping it into rivers, lakes, and the ocean -- continued in many parts of North America until the final decades of the 20th century. Clintonians probably did not give much thought to their town's sewage disposal methods, but the Water Board certainly did, and the status quo wouldn't do if the new reservoir were to live up to its billing as a source of clean water.

The problem was that sewage could no longer be dumped into the Nashua once the dam was complete, or else the expensive clean water supply would be overrun with deadly bacteria. "Continuation of the present method of disposal would seriously pollute the waters of the stream, and probably cause a nuisance. It therefore is necessary that the sewage should be diverted from the river and otherwise disposed of." Clearly, something had to be done about Clinton's sewage.

The solution was to reroute the sewage and filter the effluent through sandy soil, in southeastern Lancaster, before allowing it to return to the groundwater. Fortunately, the Water Board decided to fund this, including building a new sewer pipe connecting the old town sewer pipes with a covered, concrete structure, able to hold Clinton's unsightly waste.

Berlin, Massachusetts, had felt the coming of the reservoir in its own way. The aqueduct to Boston cut through Berlin, and the townspeople became familiar with the comings and goings of the Water Board, and the many laborers at work on the project. The Water Board supervised much of the construction from Berlin, and

Metropolitan police were in evidence in the picturesque hamlet. Although Berlin was not called upon to sacrifice any land, it lost some of its drinking wells to the aqueduct. As the aqueduct tunnel was dug below ground, it provided an alternate path for underground water supplies, causing local wells to dry up. As a stop-gap the NY, NH & H offered the water source that it used for its locomotives to the people of Berlin.

Compensation from the Water Board for all of these losses and disruptions was a sore subject for everyone from individuals to municipalities – and appeals and lawsuits were not uncommon. It may be that large industrial owners, who could afford to fight, got more attention. At any rate, the announcement of these large settlements was big news. For instance, on July 22, 1897 the Clinton Item reported:

> *A Big Transfer*
>
> *The Lancaster Mills Corporation Transfers its Sawyers Mills Holdings to Metropolitan Water Board*
>
> *One Parcel in Clinton*
>
> *The Price is Private*
>
> *Another important set of deeds was done at the office of the register of deeds in Worcester on Wednesday Afternoon, whereby the Lancaster Mills company transferred all its Sawyer's Mills property to the Metropolitan Water Board.*
>
> *This is the largest piece of property that the Metropolitan board has yet acquired...* (Daily Item).

Of course, the reservoir was a reality for everyone in its basin and the surrounding areas- living and dead. Like something out of a cheap horror film, St. Johns' Cemetery and Beaman Cemetery, were dug up, and bodies carted off. Thankfully, the bodies that were dug up were not bound for the dissection tables of any diabolic doctors. Efforts were made to make the process reasonably decorous and success was sometimes achieved. Both immigrant laborers and local residents found the process distasteful. Clinton and West Boylston appointed Martin Murphy and Charles Bray respectively to oversee the move.

Murphy and Bray were locals, and unlike the 'tramp' laborers, townspeople trusted them to respectfully move their deceased family members. However, in point of fact, most of those doing the digging were recent immigrants and others far down the wage scale. Traditions and stories preserved about these grim migrations are a fascinating side note to the vast story of the reservoir itself.

The largest exodus of the dead was certainly Clinton's St. John's Cemetery, from which close to four thousand Catholic bodies were moved. Before construction began, St. John's Cemetery (located on land currently marked as Cemetery Island, a literal stone's throw from the North Dike area of the reservoir), could be approached by a dirt road. Visitors would enter through a wooden gate in a wire fence, to pay their respects to deceased relatives. When it came time to move the dead, the whole cemetery was dug down several feet and the soil removed. This ensured that no bodies were missed and also dealt with fears that the process of disinterring the deceased might release communicable diseases. In fact, according to one source at the Clinton Historical Society, this fear of contamination made it necessary to route the remains by way of the causeway spanning Coachlace Pond (now South Meadow Pond) instead of through town.

On a warm December day, the evacuation began. Workers began to dig up bodies on each of the family plots within the cemetery, while relatives who owned the plots looked on. Most of the bodies found their way to the new cemetery just over the border of Lancaster, where they reside today at the new St. John's Cemetery. The remains of some Clintonians actually had to be moved twice. They were reburied on land once used for sewage filter beds, but due to a lawsuit, 150 bodies needed to be exhumed a second time and moved to Lancaster.

Superstitious immigrant laborers must have been on edge for the entire time they spent digging up bodies. Clinton historian, Terrance Ingano recounts a story about Italian immigrants who were digging around the roots of an elm tree, in West Boylston. The thick roots made it difficult to use shovels, and one worker pulled back a root and stood on it to keep it from springing back into place. When he stepped off the root, it sprang back like a whip, making a loud noise. The other workers panicked, and ran away from the elm tree, making the sign of the cross and offering entreaties to God. After realizing their mistake, it seems fair to say that all of the workers involved were somewhat sheepish and embarrassed.

Other, more lurid incidents occurred in the West Boylston cemetery dig. In one case, workers exhumed the bones of a baby in a box only a foot and a half below ground. There was no headstone and no cemetery record to explain the child's presence. They spoke with people who had once lived by the cemetery. Yes, they said, on one occasion they had seen a light out in the cemetery in the dead of night, but when they investigated the following morning, they had found nothing out of the ordinary. Who the child was, and why it was buried there remains unknown.

The infant's bones were not the only ones that were found buried unceremoniously. Charles Bray expected to find the remains of a colonel, a major, and the sister of the two men, based on the headstones on the plot. The sister's remains were buried in the plot, but the officers were not. Bray and his men, quite literally dug around, finally finding the bones of the two men in a soapbox, buried just within the cemetery wall. How had this come to be? Bray dug around (figuratively, this time) and discovered that the two officers had once been interred at a cemetery in Worcester. When the Boston and Worcester Railroad built an extension of its tracks, it crossed through the cemetery. It so happened, that both men were in the way, and had been moved to West Boylston. Once again, there was no explanation; this time as to why they had been buried in a soapbox rather than a coffin.

Although Edgar Allen Poe would certainly have been able to craft some lurid literature from the story of the infant and the officers, he might have been most interested in another case. Workers came across a woman, buried in a metal coffin, with a glass cover. Her body was well preserved, and the flowers she held in her lifeless hands looked quite fresh in the opinion of the workers. Supposedly, when one of them opened the coffin, they were horrified to witness the body withering within only a few minutes. The coffin was apparently hermetically sealed, explaining the woman's preserved state.

Unrelated to matters of the deceased, a scandal unfolded between 1895 and 1896, near St. John's Cemetery. Walter Brigham had purchased about a substantial amount of land in 'New Jerusalem,' Clinton's North End, while employed at a dry-goods store. He was able to turn a profit by dividing up the land into building plots, and decided to make more money by buying in south Clinton. He purchased land near the original St. John's Cemetery. This area of

Clinton was virtually empty except for the dead, and it was a long walk from the center of town. Henry Otterson, the previous owner, sold the worthless, sandy land, covered in scrub brush and scrawny oak trees, rather than pay taxes. Now that he owned the land, Brigham had plans drawn up in Worcester, to make the land appear appealing for outside investors. He planned to create a 'Western City' and mapped out streets with names such as Silver, Gold, Earle, Valley, and Grand, adding in public parks, shrubs and shade trees to heighten interest even further.

Brigham sold between 150 and 200 lots to outside investors, who had neither the time nor the curiosity to investigate Brigham's village of wonders. The scheme had worked so well, that he purchased an additional 30 acres, bringing the property to a total of 80 acres. To further 'sucker' investors out of their money, Brigham named new streets after existing streets in Clinton. Thus, if a potential investor wrote a letter to the Clinton government asking property values on (Brigham's) Beacon or Park Street, the municipal government would reply that the street was a beautiful location.

To keep the charade going, Brigham insisted on paying taxes for all of his investors. He discovered a novel way of paying off taxes each year, without dipping into his pocket. During the dry months of the year, passing locomotives would often spit out sparks or cinders that set fire to his scrub property. Every time that this happened, the railroad would pay damages, and that money went to covering investors' taxes.

When it came time to build the reservoir, Brigham's fantasy city sat atop the site for the planned North Dike. Water Board officials had to research property owners on Brigham's eighty acre plot. It was a headache for state officials, and an even greater one for investors, who finally began to realize that they had been swindled. An investor arrived in Clinton, tracked what he supposed was his house on a Clinton street bearing the same name as a street in Brigham's village, noticing angrily that someone was squatting on his property! To his chagrin and surprise, he finally discovered he was not the owner of the house in question, and in fact, could only lay claim to a plot of brambles. Some investors lost money due to Brigham's scheme, but Clintonians had never been taken in by Brigham's 'Western City;' one Clintonian commented that Brigham's land would make a good horse cemetery, and nothing else. At least the greatest swindle in Clinton history gave locals a few good laughs.

Another Clinton landmark, Cunningham's Grove, met its end as a result of the reservoir's construction. The Cunninghams owned a swath of under-developed land in south Clinton and operated the Grove, a center for functions and outings, such as dinners and picnics. Mr. Cunningham was described in 1962, by journalist Frederick Evans, as a "Character who became a legend in this area." According to Evans, Cunningham dressed like a country farmer to advertise his picnic area. "He was very large and tall (about 6 foot three or four), wore a wide brim hat and tall leather boots with pulls straps, worn on the outside of his pants or overalls with shirt sleeves rolled to the elbows, and was known as Uncle Tom." Cunningham was one of the last individuals to sell out to the Water Board. Supposedly, as the day came to leave, and for Cunningham's Grove to be cleared, gold was "discovered" on the property. Whether or not it was, we may never know, but it makes excellent material for speculation. In any case, the Water Board report for 1900 simply noted that the pavilion at Cunningham's Grove had been demolished, with the material added to the North Dike.

Some men from Clinton expected to make a different discover when Sandy Pond was drained to make way for the reservoir. As children, during a drought, they had recovered and repaired an old Native American canoe from the pond, but their parents, fearful that the boys would get up to trouble, confiscating the canoe and sank it in the middle of the pond. The boys, now grown into adults, hoped to recover the canoe and donate it to the Clinton Historical Society, but it never materialized.

While the project took much from the local towns, it sometimes left gifts in its wake. For instance, if it were not for the reservoir, Clinton might never have produced one of its most famous inhabitants. Miriam MacMillan, born Miriam Look, on June 12, 1905, was the daughter of an engineer employed on the Wachusett project. Although she did not live in Clinton for her entire life, the town can still claim connections to this polar explorer. When Miriam was just five years old she met Donald B. MacMillan, a companion of Admiral Peary and Matthew Henson on their journey to the North Pole in 1909. The year was 1911, and Donald was visiting her parents. The child Miriam was so taken with daring Mr. MacMillan that she eventually married him in 1935. Donald was 32 years older than she. Despite their age difference, the two launched 26 Arctic expeditions from their house in Provincetown. Based on her experiences in the north, Miriam wrote the novels *Etuk the Eskimo Hunter* and *Kudla and His Polar Bear*, along with a non-fiction book. Her husband died in 1970; Miriam later agreed to be the curator of Peary-MacMillan Arctic Museum at Bowdoin College, passing away at the age of 82, in 1987.

The construction of the reservoir contributed its fair share of heart-stopping surprises, too. In one instance, a 'rock crusher,' exploded, damaging houses and terrifying adults and children alike. The explosion cast large rocks into the air that shattered when they hit the ground. Few incidents were more surprising for local residents of the valley, other than perhaps the collapse of the floor, a few years later, in the Boylston Congregationalist church in the middle of a meeting that sent 100 parishioners plummeting into the basement.

One of the great ironies of the reservoir's construction was the building of the Old Stone Church by West Boylston's Baptists. Brand new in 1892, it was abandoned within only a few years of its dedication. The original Baptist church, standing on the same spot, had burned to the ground in 1890. For the sake of fire prevention, and to cement the future of Baptists in West Boylston for decades to come, the new church was sturdily built from stone, with parishioners never anticipating the plans of the Water Board. Thus, by 1895, the brand new church was already nearing retirement, its flock scattering and construction activities all around. Of all the structures within the reservoir basin, the stone edifice of the Baptist church was the only one that remained standing.

A Century of Service

The Wachusett Reservoir has had a comparatively quiet history since the tumultuous days of its construction, with a few exceptions. Unlike the Quabbin Reservoir, that had been built in the 1930s, but was still slowly filling with water during the war years, and was employed by pilots from nearby air bases, who dropped 'dud' bombs on the ruins of the Swift River valley towns, the Wachusett was hard at work serving Boston its daily supplies of water, right through the end of the war and beyond. In fact, until the Quabbin came on line in the late 1940s, most of Boston's water came from the Wachusett.

A secondary use for the reservoir emerged in 1911 four years after the reservoir had filled. Electric generating equipment was installed at the dam, meaning that the reservoir now produced power as well as water. The power house was badly damaged by a flood in 1919, when a leak sprang open, damaging records and equipment.

A large flood swept through the Connecticut and Merrimack rivers in March 1936. Factories in Lowell and Lawrence were damaged, and the Nashua River in Leominster experienced some of its highest water levels in a century. But in Clinton, the dam held and limited damage downstream.

In 1938, during the great hurricane of that year, the Wachusett spilled over its banks in places. The high winds produced by the event created large waves on the reservoir. It is unclear exactly how large these waves were, but they were large enough to wash away a section Route 110 where it is closest to the water, by the North Dike.

Sudbury Reservoir was menaced by pollution, and to ensure the continued flow of fresh water, the Hultman Aqueduct was constructed, between 1938 and 1940. The Hultman Aqueduct began at the terminus of the Wachusett Aqueduct and ended at the terminus of the Weston Aqueduct. Fully eighteen miles long, Hultman cut through three miles of solid rock beneath the Sudbury Reservoir.

After World War II, the technological achievement of the Wachusett, and its beauty, were overlooked as Bostonians focused on suburbanization, superhighways, and consumerism. Wachusett was now a stopover for the water, sent through the aqueduct from the Quabbin Reservoir, and the water from the bygone Swift River, the Quinapoxet and Nashua mingled, before rushing through another tunnel to Boston.

The up and coming '50s were a quiet time for the reservoir, robbed of tourists and day visitors by the more impressive reservoir in the west.

One of the worst droughts ever recorded in New England took place in the 1960s. From 1961 until 1969, precipitation was far below average. The Massachusetts water crisis may not be widely remembered today, but its impact was widespread. Cranberry farmers in southeastern Massachusetts had insufficient water to flood their fields, and the cranberry crop was in danger each year. The Metropolitan Water Resources Commission reported in 1965, that at least 23 municipalities had dipped into their emergency water supplies. Water bans were put into effect, but were often under-enforced, leading to a continued drop in the water supply. Civil defense officials went so far as to provide additional water pumps to eight towns. The Quabbin Reservoir, with a peak capacity of 412 billion gallons was reduced to only 45 percent of its maximum filled capacity in the mid-1960s.

With the continuing decay of Massachusetts's industry, and the expansion of highways and road trucking, the Central Massachusetts Railroad's rail line was finally closed down in 1958. Although some service persisted for longer on the western and eastern portions of the railroad, all trackage from West Berlin to Clinton Junction was taken out of service and the tracks removed.

The viaduct, however, remained because the railroad and the Metropolitan District Commission (MDC) couldn't agree about who held ownership. Finally, in 1974, the MDC decided its liability was too great and decided to have the structure dismantled. Fuller Welding Co. of Berlin was given the contract to remove all the steel work prior to June 30, 1975 – and was given the right to dispose of the steel. In fact, the trestle came down a few days late – on July 7, 1975. The anchors in the granite piers were broken, and cables were attached to the trestles. With a final tug from a bulldozer, the trestles each toppled into the Nashua, ending 62 years as Clinton's second most famous landmark.

However, the tunnel under Wilson Street remained a perfect spot for adventurous locals willing to brave its dark and musty confines. With no trains to disturb them, it also became a roosting place for bats. But one local resident, apparently expecting an imminent Cold War apocalypse, reportedly went so far as to build himself a bomb shelter within its confines. It was only when family and friends became concerned about his behavior that he was "persuaded" to leave the site – and his structure was removed. And so, the tunnel remains largely as it was, though in a few areas the roof has partly collapsed.

On the opposite end of the reservoir, the 1970s brought an end and a new beginning for the Old Stone Church. By the 1960s and '70s, the church was falling apart. Years of neglect had taken their toll. It was overgrown, its windows smashed, and souvenir hunters had removed stones from its walls, weakening its structure. There were concerns that the Old Stone Church would collapse, and it was fenced off. Daredevils managed to overcome the restrictions, wading through shallow water around the fences.

Stephen Provost's poem, reprinted in a local newspaper, captures the decrepit state of the church in the early 1970s:

I'm not very much to look at because I am so old,
But years ago, when I was young I glittered like shining gold.
I live on the banks of the reservoir and I'm usually very wet,
I've been sitting here oh so long my bricks have come unset.
Some people want to get rid of me and others they do not,
If people don't make up their minds I think I'm going to rot.

--Stephen Provost

At the beginning of the 1970s, the public once again took an interest in the church, when it was added to the National Register of Historic Places. The Beaman Oak Garden Club ran a series of pledge drives and fundraisers to support restoration, and the state legislature also voted money to its preservation. It continued to totter however, and its future was uncertain. At 12:30 a.m. in the morning, in mid-July 1974, calamity struck. The church collapsed entirely, except for two opposing walls.

"This may be a blessing in disguise. It will save money tearing it down. The stone can be kept and used to restore the building," commented Senator Harrington. Indeed, the catastrophe proved beneficial. By using money from pledge drives, the state legislature, and the Bicentennial Commission, the Old Stone Church was entirely rebuilt through 1976. The church was born anew, but lacked a roof until 1980-81. The lack of a roof did not dampen local enthusiasm. People turned out for early morning and Easter services at the mist shrouded, roofless church throughout 1979.

To celebrate the completion of restoration at the Old Stone Church, the structure was rededicated on September 25, 1983. The event was accompanied by great fanfare, including color guards assembled from Boy Scout troops, pastors, and the West Boylston High School Band.

Water, the whole point of the reservoir project, again became a matter of public focus during the 1980s, with increasing concerns nationwide about pollution. Although the reservoir has always had clean water, steps needed to be taken to ensure that clean water would continue to flow. Planners of the 19th century seemed to have assumed that population would stay stable in the reservoir towns. Likewise, few anticipated the rapid decline of the city, and the exodus of urbanites to the suburbs. Even though Boston's population dropped, Massachusetts' population continued to grow, as did the population served by the metropolitan system. The once sleepy reservoir towns were themselves experiencing population growth, and new houses began to crowd in close to the Wachusett and Quabbin reservoirs. This brought a host of new problems. Many people had septic tanks that sometimes leeched bacteria and household chemicals into the water table. Likewise, lawn fertilizer, pesticide, and other chemicals began to appear in the watershed. In 1988, reports of mercury in fish taken from the Wachusett, exceeding one part per million, led state officials to recommend that young children and pregnant women should avoid the fish.

In 1987 and 2004, complaints were heard about algae in the reservoir water that was giving drinking water a bad flavor, and contributing to odors near the reservoir. Massachusetts was criticized by the EPA for failing to do enough to keep water clean at the beginning of the 1990s. At the same time, state legislators were offered two options -- build a filtering station (that would cost between $300 million and $1.5 billion) at the Wachusett, or purchase more surrounding open land to act as a buffer against pollution. The filtration plant issue dragged on through the 1990s, but by the end of the decade, the Metropolitan District Commission, succeeded by the Massachusetts Water Resources Authority (MWRA) that manages the water supply in conjunction with the Division of Conservation and Recreation (DCR), planned to purchase 71,000 acres of land in 1996. Through 1994 and '95, the MDC took by eminent domain two thousand acres at $9,000 an acre, and an additional 15,000, between 1996 and 1998. As is often the case in the use of eminent domain, residents in the watershed areas whose land was seized often felt improperly compensated leading to three dozen law suits.

The planned filtration plant at the Wachusett reservoir may have stemmed, in part, from events in Boston. The MWRA was hard at work, addressing the overwhelming task of cleaning up Boston harbor. For decades, Boston's factories, leaking sewers, and malfunctioning harbor treatment plants had created a tremendously polluted body of water. The filthy harbor was noteworthy enough to become an argument in favor of George H.W. Bush, who pointed out that his opponent, Michael Dukakis (then governor of Massachusetts, and a pro-environment politician) presided over the most polluted harbor in the country.

The two driving concerns behind the building of the plant were the high number of the septic tanks in West Boylston and Holden that could leech bacteria into the reservoir. The other reason was the large cryptosporidium outbreak in Wisconsin. When a Wisconsin public water supply became contaminated, almost half a million people were taken ill, 4,000 hospitalized, and 100 died, in 1993. The filtration plant was originally planned to open in 1993, but was delayed.

The 1990s had highlighted pollution as a concern, for the Boston water supply system, but security became the major concern of the 2000s. Officials were at first concerned that the Wachusett Reservoir was becoming a center for suicides. In 2000, a Russian woman from Worcester leaped from the dam to her death in the reservoir. It took several days for police to find the body. Prior to 2001, the Promenade over the Wachusett Dam was open to the public. Immediately after the 9/11 attacks, the Promenade was fenced off and put under guard. If terrorists could topple two of the tallest structures in America, who was to say that they could not destroy or damage the Wachusett Dam with a truck bomb? Massachusetts Army National Guard troops were even posted around the reservoir, and walking paths were closed to public. It would be very difficult for terrorists to poison reservoir water, given the large volume of the reservoir, but state officials did not dismiss the possibility and subjected the water to a regular battery of tests. The possibility of poisoning took on new credence when two low flying planes were spotted over the Quabbin Reservoir in the immediate aftermath of 9/11, leading officials to turn off the water flow from the Quabbin.

The actions of September 2001 drastically changed attitudes about security. These concerns were, and still are well founded, as terrorism continues around the globe. It is merely unfortunate that so many new restrictions have had to be put in place, particularly on the dam, which must be an impressive structure to walk across. The 2001 security restrictions remain in effect, and will probably be around for decades to come. The closing of the Wachusett Dam to pedestrians is unlikely to be reversed, unless by citizen's action, such as a petition, or some reordering of security concerns.

Terrorism was also on the mind of West Boylston veterinarian, Dr. Robert Tashjian. Tashjian had a long running feud with state officials over botulism spores on his property. Dr. Tashjian alleged that his property had a higher than average botulism presence. State officials denied the claims, and were unimpressed with Tashjian's plan to open a biological containment center on his farm. The plan would involve the doctor's 50 miniature horses as 'canaries in the coal mine,' to detect a bio-terror attack, and would employ his farm as a quarantine center for infected animals, while working with UMass Medical School. Tashjian secured allies in the West Boylston government, starting in 2005, but local opposition and skepticism on the part of state officials make the proposal unlikely to proceed.

Between 2007 and 2008, the area around the dam was closed for a time to repair the spillway. The project involved the installation of new hydraulic gates on the spillway to control water flow. Also in 2007, officials discovered polychlorinated biphenyls (PCBs) in the soil on the Nashua River valley side of the dam. PCBs, an oil-like substance with toxic properties, were originally used in caulking the walkway across the dam in 1951. Some PCBs washed off into the soil during rain storms. Thankfully, the PCBs were localized within a few yards of the dam, and the area was fenced off. There is no apparent PCB contamination of the water, and it is not a cause for concern, because the caulking and the surrounding soil has been removed.

Efforts were under way in 2007 to eliminate the many fishing lines used by anglers, many of which were left behind. Lines left behind led to the strangulation of a muskrat, and had the potential to injure other animals.

In 2009, wet weather washed away part of the access road on the left bank of the Nashua River, just below the dam. The area was repaired. A good sign for the reservoir in 2009, was the arrival of one to two bald eagles. According to a state census of bald eagles, only 70 such birds call the Bay State home. The sighting is not the first in recent years around the reservoir; as many as three eagles were seen feeding on a dead deer a few years ago.

With the arrival of a new decade, new challenges face the reservoir and surrounding towns. Clinton suffered extensive damage from the copious rainfall in March 2010 that undermined roads, and necessitated National Guard assistance.

Of course, accidents have also been a factor in the well being of the reservoir. An unusual incident occurred in 1988, when a man, Yoshihiko Suwa, crash landed in the reservoir. Suwa took off from Sterling Airport on a solo helicopter training flight. Something went wrong with the helicopter, forcing him to ditch in the middle of the reservoir. Suwa suffered little or no injury, besides swallowing some water, and managed to swim to shore. Fortunately, the helicopter's fuel tanks did not leak, and it was salvaged within days. Another incident that could have affected water quality occurred in 2007. In early January, the driver of a Honda Civic suffered a head-on collision near the reservoir with a semi-trailer bearing diesel fuel. Much of the diesel fuel leaked out, but was quickly contained by officials. Some of the fuel leaked into a storm basin, but luckily the spill occurred five miles from the water intake, giving officials added time to diagnose the problem.

A different sort of accident befell the new Metrowest Water Supply Tunnel on May 1, 2010. A large pipe in Weston broke open, severing Boston's connection to Quabbin. Although there were no signs of contamination, a boil water order was put in place, and for a few days Bostonians and those living in outlying communities served by the MWRA snatched water bottles off of store shelves and boiled tap water. Few people in the towns receiving Quabbin and Wachusett water probably stopped to wonder why Cambridge was spared from the boil water order. In fact, the city has its own small reservoir system.

All things considered, a bright future awaits both the Wachusett Reservoir and the neighboring communities. Roman aqueducts have lasted for as much as two thousand years, and the Wachusett Dam was sturdily constructed, so it seems likely that 200, 300, even 1,000 years from now, the reservoir may still be intact, and an integral part of the Massachusetts landscape.

A Century of Service

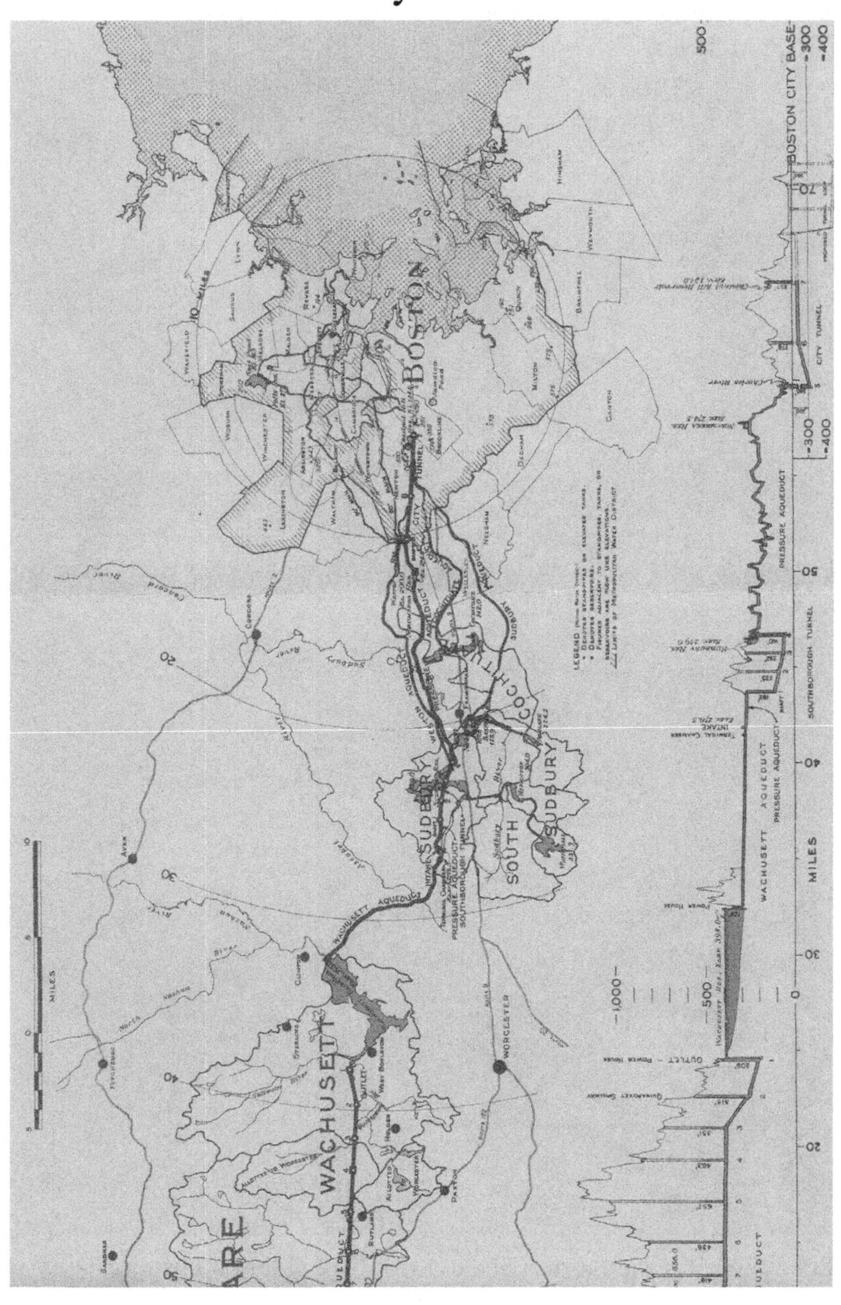

[Previous page] A map of the Boston water system extending to Wachusett Reservoir. By the time this map was produced, Quabbin Reservoir, and the Quabbin Aqueduct had been added to the system. (DCR Archives)

Even in the 1930s, Wachusett was still a technological marvel worthy of a day visit. (Grady Family)

Repair work at the North Dike in the 1950s. (Clinton Historical Society)

A snapshot overlooking the viaduct shows the spillway on a rainy day. (Grady Family)

These plastic pipes are part of a present-day initiative to safely dispose of fishing lines. (Author's Photograph)

Water diverted down a Clinton sidewalk after flooding in March 2010. (Author's Photograph)

Tailgaters Grille was put out of action temporarily during flooding in March 2010. (Author's Photograph)

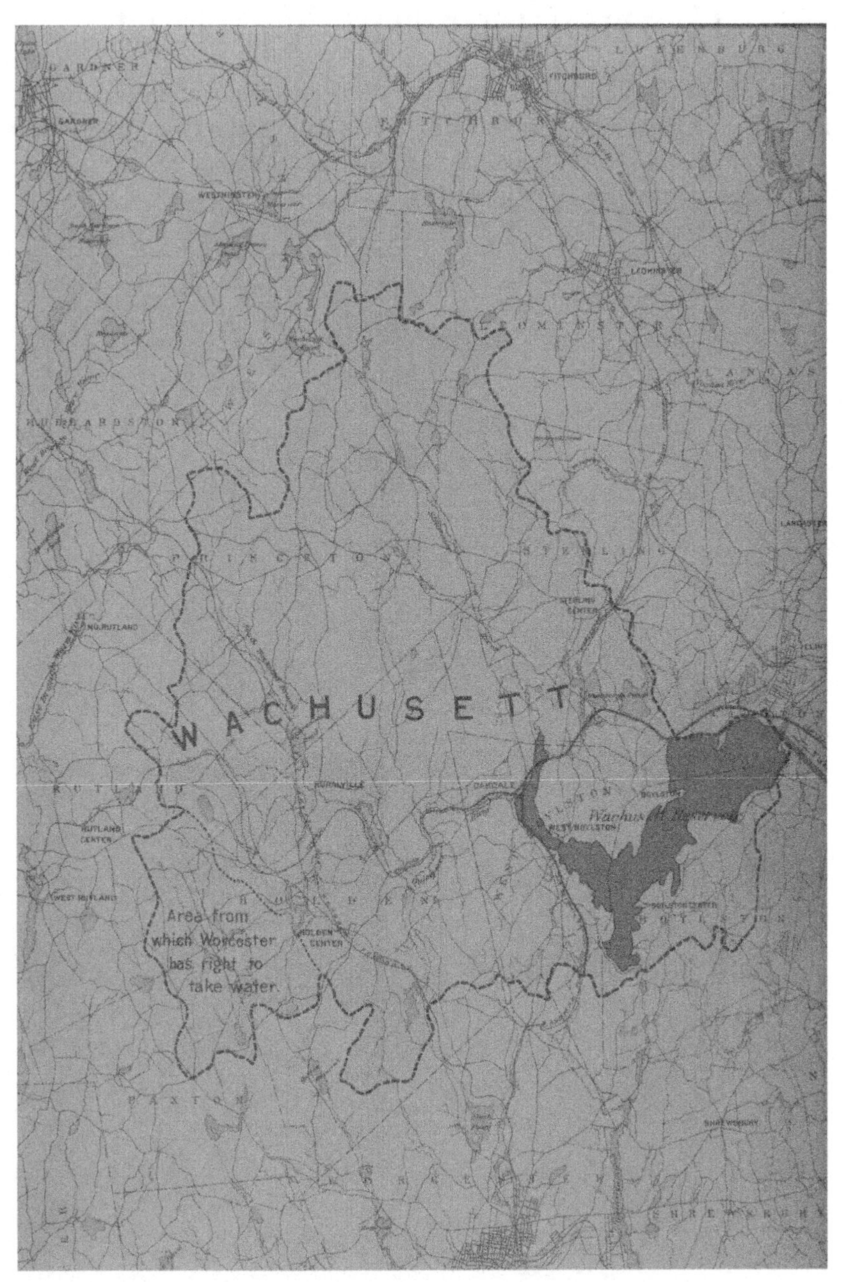

This map is a reminder that Wachusett Reservoir serves not only greater Boston, but Worcester and surrounding towns. (DCR Archive)

Removing the Viaduct

[The following still images have been extracted from a clip of 8mm film, taken by the Grady family in 1975, of the viaduct's demolition. It is the only known film of viaduct's removal.]

(Grady Family)

(Grady Family)

(Grady Family)

(Grady Family)

(Grady Family)

(Grady Family)

Ripples and waves spread across Lancaster Mill Pond, as the last component of the trestle sinks beneath the surface. (Grady Family)

All that remains of the trestle bridge: two parallel rows of stone piers. (Author's Collection)

Appendix I: Visiting Wachusett

The Wachusett Reservoir and the surrounding towns are well worth a visit and offer a cornucopia of possible activities. Within just a few miles are botanical gardens, historic movie theaters, bicycle trails, pick-your-own farms, and museums, not to mention the contiguous walking trails around the reservoir itself. Each town has something to offer.

Clinton

The brainchild of Russian icon collector, and technologist, Gordon Lankton, the **Museum of Russian Icons** is a unique cultural center in an unlikely place. The museum was formed recently, and is home to the largest collection of Russian religious icons outside of Russia itself (recently listed as one of a thousand great places in Massachusetts, by the Commonwealth). Spread out over three floors, are dozens of icons. Iconography is a religious art form, practiced by Eastern Orthodox Christianity, of which Russia has the largest number of adherents. Icons portray figures from the Bible, as well as saints, and important religious leaders from the Middle Ages and later.

Some icons date back more than 1,000 years, but despite the centuries, appear untouched by time. It is amazing that some of the icons might have been seen by figures such as Ivan the Terrible and Peter the Great who welded together Russian culture out of kaleidoscope of warring principalities, and nomadic tribes. The collection covers tremendous breadth, showing the icons from the Middle Ages to the restoration of the art form after the fall of the Soviet Union. Magnifying glasses provide an up close view of the detail in the icons. Writing too small to see with the naked eye becomes visible, written, often, centuries ago.

Gordon Lankton, the founder of the Museum of Russian Icons became interested in icons on a business related trip to Russia in the twilight of the Soviet Union. Lankton's life story sounds at times like a National Geographic article. Lankton, the chairman of Nypro, was born and raised in Peoria, Illinois. Following two years with the US Army, serving in Germany, he used his savings to purchase a motorcycle. Starting in 1956, he traveled from Europe, through the Middle East, India, Indonesia, the Philippines and Japan, before returning home. During the trip, he was transformed from a man of the Middle West to an international traveler with a new world view.

Lankton became involved with the plastics industry in Clinton, in the 1960s. It is thanks in part to Gordon Lankton, that almost all items of mass produced clothing and their labels are held together by special plastic insert tabs. The ubiquitous tab was co-invented by Lankton and his colleagues. Early experiments with plastic tabs had failed because the plastic insert was too large, leaving holes in the clothing, and breaking easily. One day, a batch of plastic tabs became overheated and stretched out. Thus the plastic tab was born. Nypro today focuses on producing high grade specialty plastics and polymers, and has branches around the globe.

Another important company located in Clinton, is British breakfast cereal company, Weetabix. The Weetabix Company, with their distribution center in Clinton is famed for their whole wheat cereal. Similar in many ways to the American brand "Wheaties" produced by General Mills, Weetabix sells its cereal in stacking, baked cakes, and generally use fewer sweeteners, than the American brand.

Many towns have seen the traditional 'downtown' fade away, as shops, stores, and restaurants move away to the outskirts of town. Clinton is an exception. Downtown Clinton is as busy as ever. One of the chief attractions in the downtown is the **Strand Theatre**. In an interesting spin on movie viewing, the proprietors have made the Strand a mix of restaurant and movie theatre. Viewers sit at tables and order meals while the film is playing. Along with playing new releases, the Strand often sets aside special times for stand up comedy routines, performed by live actors, and for "Classics," famous movies that never lose their appeal.

As of 2009, the Strand Theatre's menu includes a wide variety of food. Onion rings, curly fries, and good old fashioned french fries are available, along with chicken, hot dogs, and burgers, are available. The menu even includes veggie burgers, for the meat free crowd. If considering a visit to the Strand, a good starting place is the theatre's website (http://www.strandtheatre.com) which includes updated menus and playing times for films new and old.

For aquatic recreation, there is the **Philip Weihn Memorial Pool**, on Route 110. The Philip J. Weihn Youth Organization, is a not for profit group based in Clinton, but the pool is property of the Department of Conservation and Recreation.

Another point of particular interest for any visitor to Clinton is the **Clinton Historical Society** museum. Surprisingly, the building in which the museum is located was purpose built for the Historical Society, at the turn of the 20th century, and has never served any function other than preserving town history. Located at 210 Church Street, the museum has artifacts as varied as Civil War weaponry, Japanese samurai armor, and musical instruments. Whether or not one is interested in history, the building's architecture merits a visit. The main room is decorated with a tile mosaic floor. The building has three floors, and every minute of a visit is worth the time.

Boylston

Just southwest of Clinton is the town of Boylston. The **Boylston Historical Society** maintains a museum in a granite meeting hall leased to them by the town. The museum has a functioning bell, which is occasionally rung for special events such as holidays. The museum itself is subdivided by era. One section is devoted to displaying household items that one could purchase at the Boylston Center Store between 1900 and 1920, while another is given over to the Great Depression. The museum has an exhibit commemorating John Gough, the local lecturer and leader of the temperance movement. Many of the belongings from his house have been placed in the museum, including a samovar (a large, Russian kettle to keep tea warm) -- proof that he really was a teetotaler!

Another worthwhile visit in Boylston is **Tower Hill Botanical Garden**. One of the leading botanical groups in the state, ranking alongside Harvard Forest (in Petersham, Massachusetts) and Arnold Arboretum (in Jamaica Plain) the Garden is home to a wide variety of plants, ranging from flowers to trees. Tower Hill hosts many events every month, and is open year. Along with the plants, which are the main attraction, Tower Hill is home to a botanical library and the Twigs cafe.

West Boylston

West Boylston is located on the tapering western end of the reservoir. Like Clinton, West Boylston is home to a variety of places to visit. Perhaps most notable is the **Bigelow Tavern**, today the home of the West Boylston Historical Society's museum. The tavern has been excellently refurbished, while maintaining the feel of days gone by.

Only a short distance from the museum is **Beaman Memorial Library**. Situated at an intersection in the center of town, next to the

Masonic, the Beaman Memorial Library is great for learning more about local history, or just reading some good books. The library puts particular emphasis on children, and the entire second floor of the library is devoted to the children's' section.

West Boylston is at the same time quiet, and vibrant. It is only a few miles from Worcester, but remains a town apart.

Sterling

Sterling, Massachusetts has a number of points of interest for prospective visitors, or townies. One of the most prominent in early autumn is the **Sterling Fair** (which usually occurs in mid-September.) The town was used as a location in shooting the 2001 film *Shallow Hal* starring Jack Black. Unfortunately, one primary spot for visitation, the Mary Sawyer house, burned to the ground in 2007, possibly due to arson. **The Waushacum Ponds** are close to the reservoir, but unlike the reservoir, are open for fishing *and* boating.

Lancaster

Although it is not quite as big as Clinton, by way of towns adjoining the reservoir, Lancaster, Massachusetts is worth a visit. Unlike the other reservoir towns, Lancaster does not have a historical society which runs a museum. However, it does have a town historical commission that organizes tours of historic sites from time to time. Most often brochures for upcoming tours are published on the town website. Lacking perhaps an organized historical group, Lancaster is nonetheless steeped in history.

For whetting one's appetite, the cafeteria at the Seventh Day Adventist, **Atlantic Union College** is a great choice. Food is vegetarian, but includes pancakes, and other "comfort food." Across the street from the cafeteria, is a Seventh Day Adventist store that sells books and food. The food section in particular has a wide selection of vegetarian and health foods, many of which are organic.

Leominster

Perhaps one would not associate Leominster with the Wachusett Reservoir; however, this town is home to the **National Plastics Center and Museum**, a must-see for any visitor to the reservoir area. Leominster and Clinton are, of course, centers of the plastics industry. Due to its polymer present, Leominster was made the home of the Plastics Museum. Admission to the museum is inexpensive; exhibits are geared toward both adults and children, and examine both the history of plastics, and the cutting edge- such as plastics used in prosthetics. Whether or not one is familiar with the cutting edge or the history, one is sure to be familiar with the plastic toy building kit K'nex, which is often included in displays.

Leominster is home to **Leominster State Pool**, a public pool run by the Department of Conservation and Recreation, on Viscoloid Avenue. Also in Leominster is the **Leominster State Forest**, a 4,300 acre preserve with land in Westminster, Princeton, Fitchburg and Sterling, as well. **The Midstate Trail**, a 95 mile hiking trail crosses through the State Forest (a popular day long hike on Midstate Trail could start at **Redemption Rock** in Princeton). Boaters and anglers are free to use Paradise Pond in the state forest, and a trail is open to and maintained by the New England Mountain Bike Association.

Mt. Wachusett

When most people (other than locals) hear the word 'Wachusett' they probably think of the mountain. Located some miles west of the reservoir, the mountain is much closer than it may seem. From the summit, one can get a broad view of the landscape including (on clear days) a good view to the Wachusett Reservoir, and with a bit of luck, Boston. From the top, Boston looks a bit like the Emerald City! Located between Princeton and Westminster, Massachusetts, Mt. Wachusett is one of the taller mountains in the state, and one of the tallest in the eastern portion.

There are several trails that lead to the top of the mountain. Some are easier than others, but all of them are guaranteed to give a brief, but satisfying hike or walk. Throughout the spring and summer, the slopes are covered in an abundance of flowers, while in the winter, the ski slopes are opened. To give a longer skiing season, snow generating machines are often employed to coat the sleeps even when there is no snowfall. These generators are easily visible, driving up the paved road to the top. One can choose to drive or walk to reach the summit.

Central Massachusetts Rail Trail

An extensive rail trail network exists in the vicinity of the Wachusett Reservoir, particularly in Sterling and West Boylston. It does border directly on the reservoirs bank, but is not far away, and is excellent for biking, walking, cross country skiing, or even snowshoeing. The Central Massachusetts Railway left behind smooth pathways that have come to serve new uses in the railroad's absence.

Trails extend throughout both towns and beyond. Possibly the most well developed of them all is the roughly three mile trail leading from the very end of the reservoir in West Boylston, and following the swift flowing and beautiful Quinapoxet River. This trail in particular is well kept, with several new steel bridges crossing the Quinapoxet. Midway down the trail, one can find the remains of a mill. Overgrown as it is, each position within the old mill's foundation is marked with a granite pylon and new signs. An unusual monument is located at the beginning of the trail, where two green cars have been laid down, one vertically and the other horizontally, forming an 'L' shape. The monument is meant to remind pedestrians and bicyclists of the line that once ran through here. Although this is one of the main stretches, the Central Massachusetts Rail Trail branches out around the reservoir offering extensive paths for recreation

Safety and Regulations

So far, mention has been made only of the reservoir towns, and outlying areas that somehow have a connection to the reservoir. All of these places are worthwhile visits, but so is the reservoir itself. Almost all of the area immediately surrounding the reservoir is owned by the Commonwealth of Massachusetts, and is managed by the Department of Conservation and Recreation (DCR). The DCR is charged with both protecting the reservoir from pollution, and keeping land in shape for public recreation.

It is important to keep in mind, while exploring around the reservoir, the many rules that have been put in place to protect the water supply and the public. The Wachusett Reservoir is not open for boating (although one may spot boats belonging to the government at times). The reservoir is open for fishing however, and there are numerous gates located at different points around the reservoir, where one can fish from shore. Gates where fishing is allowed include gates 6 to 16, on Rt. 70, 17 to 24 on Rt. 140, and 25 to 35 on Rt. 12/110, and lastly the Thomas Basin in West Boylston. Before planning a fishing visit it is also a good idea to visit the DCR website (http://www.mass.gov/dcr/parks/central/wachRes.htm) where the latest updates are posted (including water qualities reports that may relate to fishing). The fishing season runs from the first Saturday in

April (so long as there is no ice present) until November 30th, from dawn to dusk each day.

The government has a well developed sign system that should be easy for the public to understand. Certain gates and parts of the shoreline have signs that warn against trespassing. These signs are not universal, and by driving, walking, or biking further west along the south shore, one will encounter several unpaved parking spots located at the start of trails and paths that are open to the public. Do not trespass, because the rangers really do patrol the reservoir and will arrest violators! A good rule- if you are ever in doubt, about whether or not you are entering a no trespassing zone (some are vaguely marked, so this is possible) do not go any further, and head back to a place where the public are meant to stay.

Some areas around the reservoir may be closed at times for renovation. The DCR has the latest updates on any closures or new regulations. The following are some general ground rules to keep in mind. Walking and hiking are permitted in all places without no trespassing signs. Boating (except for non-motorized boats on Waushacum Pond), swimming, ice fishing or ice skating are prohibited as well as snowmobiling and off-roading. Building fires, cutting new paths, using metal detectors and horseback riding are forbidden. The good news- picnicking is permitted at Waushacum, and certain areas such as the **Old Stone Church**. Hunting is also permitted but only in headwaters zone (beyond the western end of the reservoir). Hunters can use hunting dogs in this area. Cross country skiing is also permitted in some areas.

Signs around the reservoir also warn pedestrians not to be alarmed by the sound of gunfire. Loud sounds such as the sound of gunfire are employed to scare birds away from the reservoir, where their droppings might cause bacterial pollution of the drinking water. Around the dam, be aware of sudden water rise. Sometimes during the spring, or in rainy weather, the spillway will rapidly fill with water, leading to a rise in the Nashua River's level. If this is the case, sirens will sound, and one can climb the slopes of the valley on either side to reach higher ground. To police the reservoir, the DCR has a group called the Watershed Rangers (978-365-3800) which keeps an eye out for polluters and trespassers. If you ever spot anything out of the ordinary, the State Police barracks in Holden (508-829-8410) is also charged with protecting the reservoir.

Be respectful of the DCR Watershed Rangers. They are here to do an important and underappreciated job, protecting the public water supply. Besides policing the reservoir, the Watershed Rangers offer frequent tours and programs for the public relating to the reservoir's history. These tours examine the buildings and landscapes that existed before the reservoir, and how it was constructed. Remember when exploring the reservoir to follow regulations, have a good time, stay safe, and most of all do not pollute -- the water in the reservoir is drinking water for almost a million people!

Visiting

The Old Stone Church in West Boylston, one of the only reminders of West Boylston's submerged neighborhoods. (Author's Photograph)

An unusual monument to the Central Mass. Railroad in West Boylston. (Author's Photograph)

Downtown Clinton, showing from left to right the Holder Memorial Museum, Bigelow Free Public Library, Clinton's Town Hall, and the Seventh Day Adventist church. (Author's Photograph)

Strand Theatre takes an interesting approach to movie going, combining a restaurant and a cinema. (Author's Photograph)

Brain-child of Clinton philanthropist Gordon Lankton, the Museum of Russian Icons is the largest collection of religious icons outside of Russia. (Author's Photograph)

Boylston Public Library, seen here from across the common, is a focal point of the community as well as a unique building and excellent library. (Author's Photograph)

The former town hall, houses the Boylston Historical Society Museum, seen here, from across the common. (Author's Photograph)

The Beaman Memorial Public Library is both an excellent library, and a center of West Boylston life (Author's Photograph)

The Quinapoxet River flows into the Wachusett Reservoir in West Boylston (Ethan McCarthy Earls)

The Quinapoxet Dam, 'Quinnie,' seen from the Central Mass. Rail Trail. (Author's Photograph)

A CSX train carrying electric generating equipment crosses the Thomson Basin in West Boylston, one of the best fishing areas on the reservoir. (Author's Photograph)

West Waushacum Pond seen from the Central Mass. Rail Trail. The pond is used for fishing and boating. (Ethan McCarthy Earls)

The Quag is a man made pond, created from West Waushacum Pond by the railroad. The two ponds are still connected by a narrow channel that is a popular spot for fishing. (Ethan McCarthy Earls)

Shown in the off-season, devoid of snow, the ski slopes at Mt. Wachusett make excellent skiing during the winter, and are only a stone's throw from Worcester and Boston. (Author's Photograph)

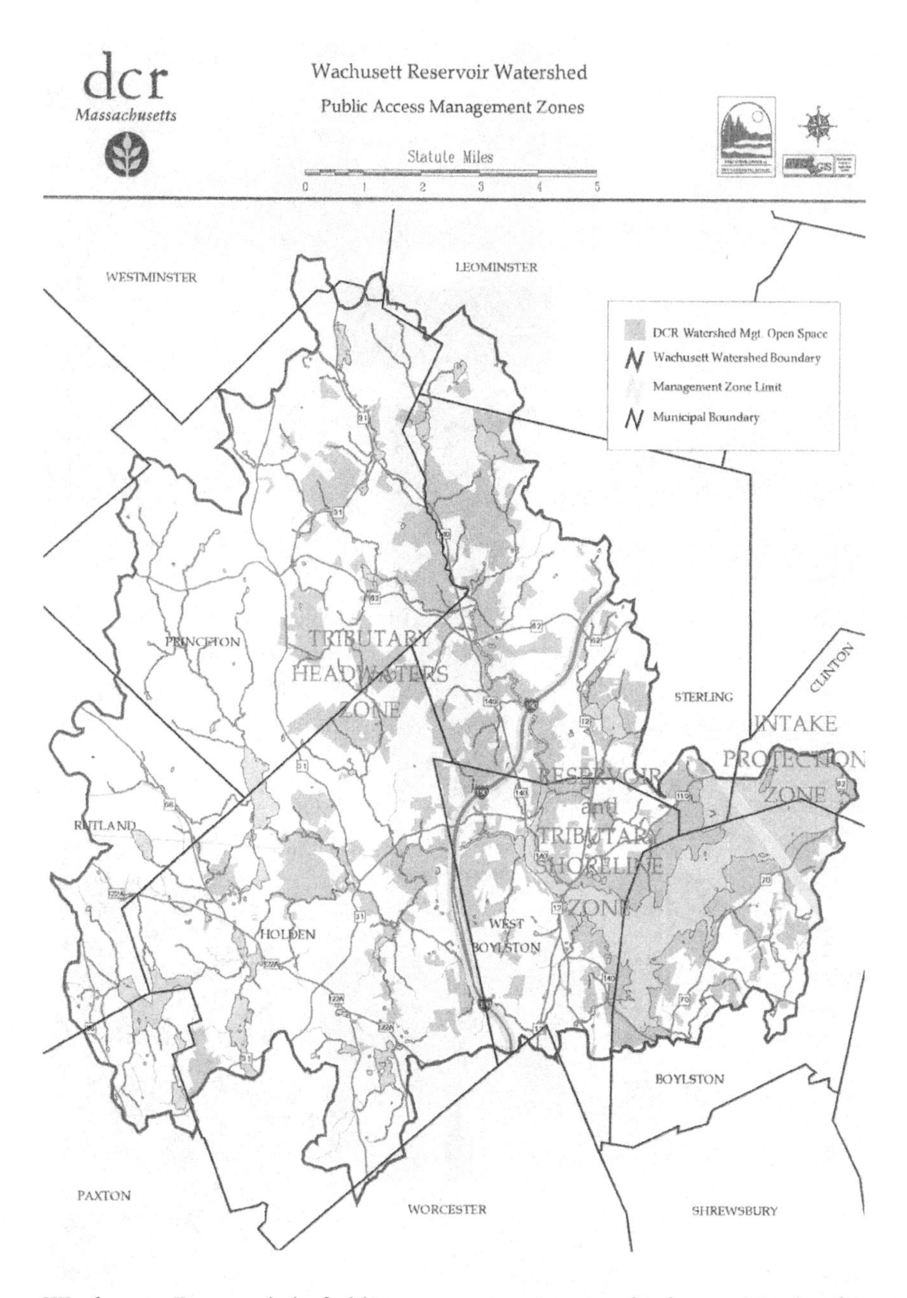

Wachusett Reservoir is fed by an enormous watershed area. Much of the land in the watershed is conserved and protected by the DCR Watershed Rangers. (Department of Conservation and Recreation)

A map of the reservoir showing access points for recreation.

(Department of Conservation and Recreation)

Appendix II:
Water for Worcester and Marlborough

With 182,000 and 38,000 people respectively, Worcester and Marlborough are both close neighbors of the Wachusett Reservoir. Of course, Wachusett was a Boston-led project, but it could just as well have been a project launched by the City of Worcester. The second largest city in Massachusetts had its own, small, water expansion, along with Marlborough (or Marlboro; the spelling battle has yet to be won.) By looking at these smaller water systems, it is easier to understand the immense Boston water project.

Worcester faced many similar challenges to those faced by Boston- a mushrooming population, mills, and factories. In 1845, during the same decade that Boston expanded its water supply by raising the level of Long Pond; the Worcester Aqueduct Company (WAC) took control of Bell Pond. Three years later, the city purchased the WAC. In 1864, Mayor D. Waldo Lincoln added a dam on Lynde Brook. The new reservoir held 681 million gallons. Later on, in 1883, the city built the Tatnuck Brook Dam, adding 730 million gallons to the city supply.

Tatnuck Brook Dam came about after a campaign to introduce water meters in Worcester. The introduction of water meters saved the city hundreds of thousands of gallons of water, but it still needed more. 1898 Worcester had 29,256, sinks, 2,237 stables, 13,592 baths – and 98 saloons, whose water needs were supplied by the city system. However, Tatnuck Brook would not be the last.

One of the most recent additions to the Worcester system is the small Quinapoxet Reservoir, completed in 1953. Worcester has never had projects as grandiose as those masterminded by Bostonians, and they have tapped into the same system at times, drawing water out of the Wachusett Reservoir in emergencies (part of the reason why Worcester was not forced to build more reservoirs). Worcester, nonetheless, has had its share of water concerns, including freshwater shrimp that moved into its water pipes!

Marlborough has a somewhat smaller system than Worcester. Geographically it spills over into Northborough and Hudson. Fort Meadow Reservoir, today overrun with Asiatic clams, was built in 1848. At one time, Fort Meadow had been a swamp, through which flowed Fort Meadow Brook, feeding into the Assabet River. The Maynard family dammed up the brook to power their grist mill, eventually selling water rights to Boston in 1847, and the pond was enlarged. A decade later, Boston decided that it no longer needed Fort Meadow Reservoir, and sold it back to the Maynard family. It remained in service as a water supply until the end of the century. Once it was taken out of service, it became a local swimming hole, and cottages were built on its shores. In 1953, Marlborough bought waterfront for a town beach.

Fort Meadow Reservoir is not the only reservoir in Marlborough. Sudbury Reservoir extends into the south of the town, and the town draws drinking water from Millham Reservoir and Lake Williams. Lake Williams is located near I-495, and skirted by Lakeside Avenue. Constructed in 1893, is another town supply, Millham Reservoir. Adjacent to Millham Reservoir, but unrelated, is Tyler Dam, a concrete flood control dam, alongside Robin Hill Street.

In telling the story of the Wachusett Reservoir, it is important to give some passing mention to these two systems, as points of reference for Boston's bigger project. Even medium size towns such as Marlborough, have at one time embarked on a reservoir building spree. Perhaps reservoirs are a measure of a city or town's importance, or at least how it sees itself in comparison with its neighbors. Massachusetts is a patchwork of overlapping reservoirs, watersheds, aqueducts and municipal domains that extends to Bear Swamp Hydroelectric Power Station in Florida, Massachusetts. These domains feed into one another, and the water claimed by one city may find its way into the drinking glass of another before long.

Bibliography

•"State Warns on Mercury Levels of Fish in Wachusett Reservoir." The Boston Globe. 30 March, 1988.

•A General Description of the Water Supply of the Boston Metropolitan District. Commonwealth of Massachusetts; 1940.

•Allen, Scott. "Mass. Sees win in war with gulls Wachusett Reservoir cleaner officials say." The Boston Globe. 23 March, 1995.

•Bastarache, A.J. An Extraordinary Town: How One of the Smallest Towns Shaped the World. Clinton; Angus MacGregor Books. 2005.

•Clinton Daily Item. 1893, 1895-1908.

•Colangelo, Matthew P. The Wachusett Reservoir and the Town of West Boylston. Worcester Polytechnic Institute. 1984.

•Cook, Gareth; Daley, Beth. "Vigilance Grows Over Water Sept. 11 Attacks Prompt Tighter Security at State's Reservoirs." The Boston Globe. 26 September, 2001.

•Curran, Karen. "The Great Wachusett Landgrab the State Wants 71,000 Acres Protected; Property Owners Want Fair Market Value." The Boston Globe. 14 January, 1996.

•Dumanoski, Dianne. "EPA Tells State: Protect Reservoirs or Build Water Filtration System." Boston Globe. 21 November, 1990.

•Dumanoski, Dianne. "MWRA Postpones Wachusett Filtration Plant." Boston Globe. 10 June, 1993.

•Gottesman, Jan. "Eagles Make Reservoirs Home." Worcester Telegram & Gazette. 29 January, 2009.

•Greene, J.R. The Creation of the Quabbin Reservoir: The Death of the Swift River Valley. Athol. The Transcript Press. 1981.

•Heslam, Jessica. "Body Found in Reservoir May be Worcester Woman." Boston Herald. 25 July, 2000.

•Hopfmann, Ruth M. The Wachusett Reservoir. Third Annual Local History Conference Montachusett Region. 1983.

•Ingano, Terrance. Twice Told Tales of Clinton. Clinton; Angus MacGregor Books. 2007.

•Ingano, Terrance. Twice Told Tales of Clinton 2. Clinton; Angus MacGregor Books. 2009.

•James, Patricia J. "Diesel fuel spills into reservoir; Wachusett Reservoir being cleaned up after accident." Worcester Telegram & Gazette. 5 January, 2007.

•Lepore, Jill. Resistance, Reform, and Repression: Italian Immigrant Laborers in Clinton, Massachusetts, 1896-1906. Harvard University. 1989.

•MHC Reconnaissance Town Survey Report. "Boylston." 1983.

•MHC Reconnaissance Town Survey Report. "Clinton." 1983.

•MHC Reconnaissance Town Survey Report. "Sterling." 1983.

•MHC Reconnaissance Town Survey Report. "West Boylston." 1983.

•Magiera, Anne Mary. "'Quinie Dam May be Falling.'" Worcester Telegram & Gazette. 19 August, 2009.

•Magiera, Mary Anne. "Wachusett cleans up line." Telegram & Gazette. Worcester Telegram & Gazette. 27 July, 2007.

•Nugent, Karen. "Toxins found in soil at dam; PCBs detected at Wachusett Dam from walkway caulking." Worcester Telegram & Gazette. 23 March, 2007.

•Olson, Scott M., Stark, Timothy D., Walton, William H., and Castro, Gonzalo. "1907 Static Liquefaction Flow Failure of the North Dike of the Wachusett Dam." Journal of Geotechnical and Geoenvironmental Engineering. December 2000.

•Ouelette, Hank. "A Brief History of the Springdale Mill." Wachusett Greenways. Web. 3 May 2010.

•Pratt, Robert M. "The Earth Work of the Wachusett Reservoir." Water and Sewage Works. 1902.

•Report (of the Metropolitan District Commission) to the Massachusetts Legislature, 1896. Wright and Potter; 1896.

•Report Upon the Value of Albertson's Water Power and Property at Boylston, Mass. Dean & Maine Mechanical & Mill Engineers. 1899

•Semi-Centennial Celebration of the Incorporation of the Town of Clinton: Mar. 14 1850; June 17-18-19 1900. Town of Clinton. 1900.

•The Central Mass. Boston. Boston & Maine Railroad Historical Society, Inc. 1975.

•Towns of the Nashaway Plantation. Hudson; Lancaster League of Historical Societies. 1976.

•Upano, Alicia. "U. Massachusetts Area Reservoir Remains Closed After Suspicious Aircraft Spotted Sept. 22." University Wire. 1 October, 2001.

•Van Sack, Jessica. "Bad Taste in Water Traced to Algae in Reservoir." Patriot Ledger. 10 July, 2004.

•Wandle, Jr., William S. "Massachusetts: Floods and Droughts." National Water Summary 1988-89- Floods and Droughts: Massachusetts.

•Water Supply and the Work of the Metropolitan Water District. Boston; Metropolitan Water District. 1900.

•Weeks, John. "Wet and Wild; State Tests Dam Spillway Doors." Worcester Telegram & Gazette. 11 July, 2008.

•"8 Tunnel Workers Hurt in Mass." AP Online. 11 May, 2000.

•"Man Swims Ashore After Helicopter Falls in Reservoir." Boston Globe. 19 June, 1988.

•"State Takes Action to End Environmental Impact to Wachusett Reservoir by Wachusett Regional High School Site." US Fed News Service. 21 February, 2006.

•"Wachusett Reservoir- 100 Years Old." Downstream. Department of Conservation and Recreation; 2005.

•"Wachusett and Sudbury Reservoir Fishing Guide." 2009. Department of Conservation and Recreation. 14 February 2010.

Index

About the Author

Eamon McCarthy Earls first became interested in the Wachusett Reservoir and the Boston water system, after visiting the Wachusett Dam in 2004. After reading many of the books written about the Quabbin Reservoir and the Boston water system as a whole, he began to seek information about Wachusett. To his surprise, he found no single work existed on the subject, leading him to begin the research which has led to the present volume. Research began in late 2008, and writing was finally completed in late spring 2010. This is his second book. *Kearns On the Double* a mystery novel, comprising two stories set in a fictitious Massachusetts town during the 1920s and '30s, was published in 2009.

Made in United States
North Haven, CT
14 August 2022

22716463R00088